A New Approach to the Nuclear Fuel Cycle

Best Practices for Security, Nonproliferation, and Sustainable Nuclear Energy

AUTHORS
Kelsey Hartigan
Corey Hinderstein
Andrew Newman, PhD
Sharon Squassoni

Foreword by former Senator Sam Nunn

February 2015

A Report of the CSIS Proliferation Prevention Program and the Nuclear Threat Initiative

CSIS | CENTER FOR STRATEGIC & INTERNATIONAL STUDIES

NTI BUILDING A SAFER WORLD

ROWMAN & LITTLEFIELD
Lanham • Boulder • New York • Toronto • London

ISBN: 978-1-4422-4053-7 (pb); 978-1-4422-4054-4 (eBook)

CREDITS

Cover photo: Photo by URENCO.
Page 6: Argonne National Laboratory, "Experimental Breeder Reactor II," October 2, 2002, http://commons.wikimedia.org/wiki/File:EBRII1.jpg.
Page 9: International Atomic Energy Agency (IAEA), "1366th Board of Governors Meeting," Vienna, November 28, 2013. Photo credit: Dean Calma/IAEA, https://www.flickr.com/photos/iaea image-bank/11098864516/.
Page 17: Nuclear Regulatory Commission, "The entrance to the proposed Yucca Mountain high-level waste repository site under construction in 2007," August 12, 2007, http://www.nrc.gov/reading-rm/doc-collections/fact-sheets/yucca-license-review.html.
Page 18: Lawrence Livermore National Laboratory, "Atomic Vapor Laser Isotope Separation (AVLIS)," January 6, 2012, http://commons .wikimedia.org/wiki/File:AVLIS laser.jpg.
Page 19: National Nuclear Security Administration, "An aerial photo shows the full extent of construction at the MOX Project in mid-2011," August 5, 2011, https://www.flickr.com/photos/nnsanews/6012112070/in/photolist-aagDsE-aadQq8-9zBuKi-8ScAxN-8pe3EA-8S9u7B-8S9uqK-aadQtp.
Page 28: Department of Energy, "A Transuranic Package Transporter (TRUPACT-II) entering the Waste Isolation Pilot Plant (WIPP) outside of Carlsbad, NM," June 12, 2014, https://www.flickr.com/photos/departmentofenergy/14384113456/.
Page 31: Nuclear Regulatory Commission, "Spent fuel dry casks at a nuclear power plant site," November 15, 2007, https://www.flickr.com/photos/nrcgov/6946374745/.
Page 37: Japan Nuclear Fuel Limited, "The Rokkasho Reprocessing Plant in Japan," n.d., http://www.jnfl.co.jp/english/business/reprocessing.html.
Page 50: World Nuclear Association, "The Nuclear Fuel Cycle," October 2014, http://www.world-nuclear.org/info/Nuclear-Fuel-Cycle/Introduction/Nuclear-Fuel-Cycle-Overview/.
Page 51: IAEA, "Spent Fuel Pool," November 27, 2013. Photo credit: Greg Webb/IAEA, https://www.flickr.com/photos/iaea imagebank/11083570446/.
Page 54: Department of Energy, "The Three-Drum Filtering Process for Extracting Uranium Oxide from Uranium Ore," September 24, 2014, https://www.flickr.com/photos/departmentofenergy/15157631999/.

Center for Strategic & International Studies
1616 Rhode Island Avenue NW
Washington, DC 20036
202-887-0200 | **www.csis.org**

Nuclear Threat Initiative
1747 Pennsylvania Avenue, NW, Seventh Floor
Washington, DC 20006
202-296-4810 | **www.nti.org**

Rowman & Littlefield
4501 Forbes Boulevard
Lanham, MD 20706
301- 459- 3366 | **www.rowman.com**

TABLE OF CONTENTS

TABLES AND FIGURES

ACKNOWLEDGMENTS

This project would not have been possible without the contributions of many. Very special thanks are due to the John D. and Catherine T. MacArthur Foundation for its generous support of the work.

We are extremely grateful for the advice and input of our "core group" members whom we consulted throughout our efforts. While they helped shape our thinking, they are not responsible for the final content of the report. These individuals include John Carlson, Charles B. Curtis, Alan Hanson, Angie Howard, Tom Isaacs, Fred McGoldrick, C. J. Milmoe, Bob Rosner, Tom Sanders, Larry Scheinman, Carl Stoiber, Melissa Mann, and Emma Belcher.

We also want to thank all the professionals who gave time and effort to participate in project workshops, breakout sessions, and red-teaming efforts. A special thanks goes to Stephanie Cooke for her work on nuclear industry trends.

At CSIS, additional support was provided by Tamara Spitzer-Hobeika, Leah Fae Cochran, and Bobby Kim. At NTI, we were also supported by Lauren Callahan and Herbert J. Scoville Fellow Jessica Bufford.

Finally, we are indebted to the leadership and Boards of Directors of both NTI and CSIS. In particular, we thank President Joan Rohlfing and Executive Vice President Deborah Rosenblum of NTI for their strategic vision and feedback.

COREY HINDERSTEIN AND SHARON SQUASSONI
Project Directors

ABOUT THE NEW APPROACHES TO THE FUEL CYCLE PROJECT

In the past decade, a resurgence of enthusiasm for nuclear power has rekindled interest in efforts to manage the fuel cycle. The 2011 accident at the Fukushima Daiichi nuclear power plants in Japan and current proliferation crises in North Korea and Iran raise this question: Is the current approach on the fuel cycle—leaving uranium enrichment and spent fuel reprocessing capabilities in the hands of national governments—too risky on proliferation grounds? New approaches to the nuclear fuel cycle with the objective of mitigating proliferation risks can also help improve nuclear governance, making nuclear energy safer and more sustainable.

In early 2011, the Nuclear Threat Initiative and the Center for Strategic and International Studies launched the New Approaches to the Fuel Cycle (NAFC) project. This project, led by Corey Hinderstein and Sharon Squassoni, sought to build consensus on common goals, address practical challenges, and engage a spectrum of actors that influence policymaking regarding the nuclear fuel cycle. Drawing from industry, government, and NGO community expertise in the United States and abroad, the NAFC project worked to outline a vision for an integrated approach to nuclear supply and demand. The project, which hosted multiple workshops and smaller breakout groups to vet ideas, sought specifically to identify practical solutions that could be adopted in phases.

The result is the first comprehensive approach that contains guidelines for shaping a sustainable nuclear supply system and leverages existing trends in nuclear industry.

This approach offers a set of "best practices" to help implement that sustainable system. The project also tackled one of the toughest issues—spent nuclear fuel and high-level waste—to see if solutions there might offer incentives to states on the front end of the nuclear fuel cycle and address the inherent inertia and concerns about additional burdens and restrictions that have stalled past efforts to improve the robustness of the nonproliferation regime. This report presents the group's conclusions that a best practices approach to the nuclear fuel cycle can achieve these objectives and offer a path to a more secure and sustainable nuclear landscape.

FOREWORD

IN 1946, ALBERT EINSTEIN SENT A TELEGRAM TO SEVERAL HUNDRED PROMINENT AMERICANS ASKING THEM TO SUPPORT A NATIONAL EDUCATION CAMPAIGN ON THE GRAVE DANGERS POSED BY THE ATOM. "The unleashed power of the atom has changed everything save our modes of thinking, and we thus drift toward unparalleled catastrophe," the physicist warned.

His words resonate today, nearly 70 years later, as parties prepare to gather in 2015 for the conference held every five years to review implementation of the Treaty on the Non-Proliferation of Nuclear Weapons (NPT).

The NPT, signed by 189 countries and backed by every U.S. president since Lyndon Johnson, has enshrined the obligation of working toward a world without nuclear weapons. Since the treaty entered into force in 1972, efforts to meet the implicit bargain it struck—that recognized nuclear states would take steps to review and eliminate nuclear weapons and states without nuclear weapons would not acquire them—have undoubtedly made the world safer.

However, the NPT has two major flaws: It offers a vision of disarmament and nonproliferation without a coherent plan or benchmarks to get there, and it places no restrictions on states' rights to nuclear technology for peaceful purposes.

To be fair, when the treaty was negotiated, it was widely believed that the ability to acquire fuel cycle technology and know-how was beyond the reach of all but a few countries. Treaty drafters likely could not imagine a day when enrichment technology would be illicitly transferred between states or when terrorist organizations bent on massive destruction would be seeking nuclear materials.

Unfortunately, that is where we are now. Today, it is clear that sensitive information related to the nuclear fuel cycle cannot always be protected or contained. The claim under the NPT of a "sovereign right" to nuclear technology for peaceful purposes has led to the proliferation of technology that also can be used to produce weapons-usable nuclear material, giving states a latent nuclear-weapons capability.

Everyone understands that there's no putting the genie back in the bottle when it comes to the spread of nuclear science. Most also agree that the peaceful applications of nuclear science can play an indispensible role in our efforts to meet human needs in the twenty-first century, including for our energy and environmental future.

However, because the same materials that are used to fuel power plants, if further enriched, can be used to build weapons that can destroy the planet, we must do more to prevent the catastrophic use of the atom. The promise of our nuclear future depends on how we manage our nuclear present.

So managing the fuel cycle—at a time when as many as three dozen countries are reportedly interested in building their first nuclear power plant—is paramount. To address the practical challenges involved and engage those who can influence policymakers, the Nuclear Threat Initiative (NTI) and the Center for Strategic and International Studies (CSIS) joined forces in 2011 on a two-year project that explored ways to mitigate security and proliferation risks while developing principles for a secure and sustainable nuclear fuel cycle architecture.

The results are contained in this report: a sensible "best practices" approach that builds on industry trends in corporate responsibility and industry globalization while offering a path to a more secure future.

Unfortunately, the world so far has been reluctant to confront the key question raised by security gaps in the NPT: Do we really believe that we can live securely in a system that poses so few constraints on any state's ability to produce weapons-usable nuclear materials?

I believe that the answer is no. From the continued buildup of fissile materials around the world to the lack of will and capacity to safely dispose of spent nuclear fuel, the current trajectory is unacceptable. It is past time for leaders to tackle the difficult issues posed by today's fuel cycle.

As Einstein said, it is time to change our thinking. We believe this report provides a roadmap for leaders to reconsider the current path and take the urgent action needed to create a safer world.

SAM NUNN
Co-Chairman and Chief Executive Officer
Nuclear Threat Initiative

Chairman, Board of Trustees
Center for Strategic and International Studies

EXECUTIVE SUMMARY

WITH THE HINDSIGHT OF 60 YEARS, POLICYMAKERS TODAY WOULD LIKELY DESIGN A GLOBAL NUCLEAR FUEL CYCLE ARCHITECTURE QUITE DIFFERENT FROM THE ONE THAT HAS EVOLVED. That architecture would support nuclear energy development but also ensure that nuclear materials and technology were properly secured. It would be structured to satisfy the needs of all states, whether they have nuclear weapons or not, and whether they are nuclear suppliers or customers, now or in the future.

It's too late to design a fuel cycle architecture from scratch, but it's not too late to refine the fuel cycle so that it reflects our common objectives of protecting against the spread of nuclear weapons to new countries and against catastrophic acts of terrorism with nuclear materials, while also protecting the long-term viability of nuclear energy to provide low-carbon electricity. To facilitate new approaches, all stakeholders—governments, industry, regulators, and other experts—must cooperate to design and implement solutions.

The new approach to the fuel cycle outlined here is the result of a two-year collaboration between the Nuclear Threat Initiative (NTI) and the Center for Strategic and International Studies (CSIS) that drew on U.S. and foreign expertise in the nuclear industry, government, and the nongovernmental organization (NGO) community. It is different from past approaches in three ways: It is comprehensive (including the front and back end of the fuel cycle), it offers broadly acceptable guidelines ("best practices"), and it leverages existing market trends. Unlike past proposals, it directly addresses the fundamental, underlying issues of sovereignty and uneven distribution of capabilities and opportunities within the nuclear fuel cycle.

The best practices advocated in this report constitute the first comprehensive approach to creating a sustainable nuclear supply architecture. It is sustainable because it manages risks and rewards for compliance. It is comprehensive because it addresses both front and back ends of the nuclear fuel cycle—that is, the production, use, and disposal of fissile material and nuclear waste. At the same time, it targets both nuclear security and nonproliferation risks and distributes responsibilities among nuclear weapon states and non-nuclear-weapon

Agreement on fundamental, international security objectives can help overcome the inertia that has been a primary stumbling block in this policy arena for decades.

states. It addresses not just reducing current risks but also advocates consequences for noncompliance with norms.

The best practices help establish norms of operation in the system: equity, shared costs, early detection of noncompliance, minimization of weapons-usable material, market-driven expansion of capabilities, and balanced incentives and disincentives. These practices suggest that agreement on fundamental, international security objectives can help overcome the inertia that has been a primary stumbling block in this policy arena for decades. Therefore, the guidelines are an important interim step in identifying solutions.

Finally, the approach outlined in this report seeks to leverage existing market trends. Rather than imposing a new system of control upon industry, the action plan that is derived from the guidelines builds upon existing trends in industry globalization and corporate responsibility. The report identifies specific projects for government, industry, and civil society collaboration that will help fine-tune the contours of a more secure and sustainable nuclear landscape.

The Risks

The continued buildup of fissile material stockpiles, the spread of uranium enrichment and spent fuel reprocessing facilities, and the lack of will or capacity to dispose of spent nuclear fuel all pose long-term risks for proliferation and nuclear security. For a global community that recognizes both the value and risks of nuclear energy, this trajectory is unacceptable.

The spread of technologies to produce nuclear fuel for reactors—uranium enrichment and spent fuel reprocessing—is perhaps the most challenging of those dangers. These sensitive technologies can be used to make benign fuel or nuclear explosives. Because fissile material production is often the biggest hurdle in a nuclear weapons program, having a ready-to-go capability poses real security and proliferation risks.

In theory, any state could engage in all the activities listed below while being in full compliance with the Nuclear Non-Proliferation Treaty (NPT). These activities ultimately could aid a nuclear weapons program:

- Producing highly enriched uranium (HEU)
- Withdrawing HEU from safeguards for nonexplosive military purposes
- Accumulating stocks of separated plutonium or low-enriched uranium (LEU)

Figure ES-1. Amounts of Weapons-Usable Material

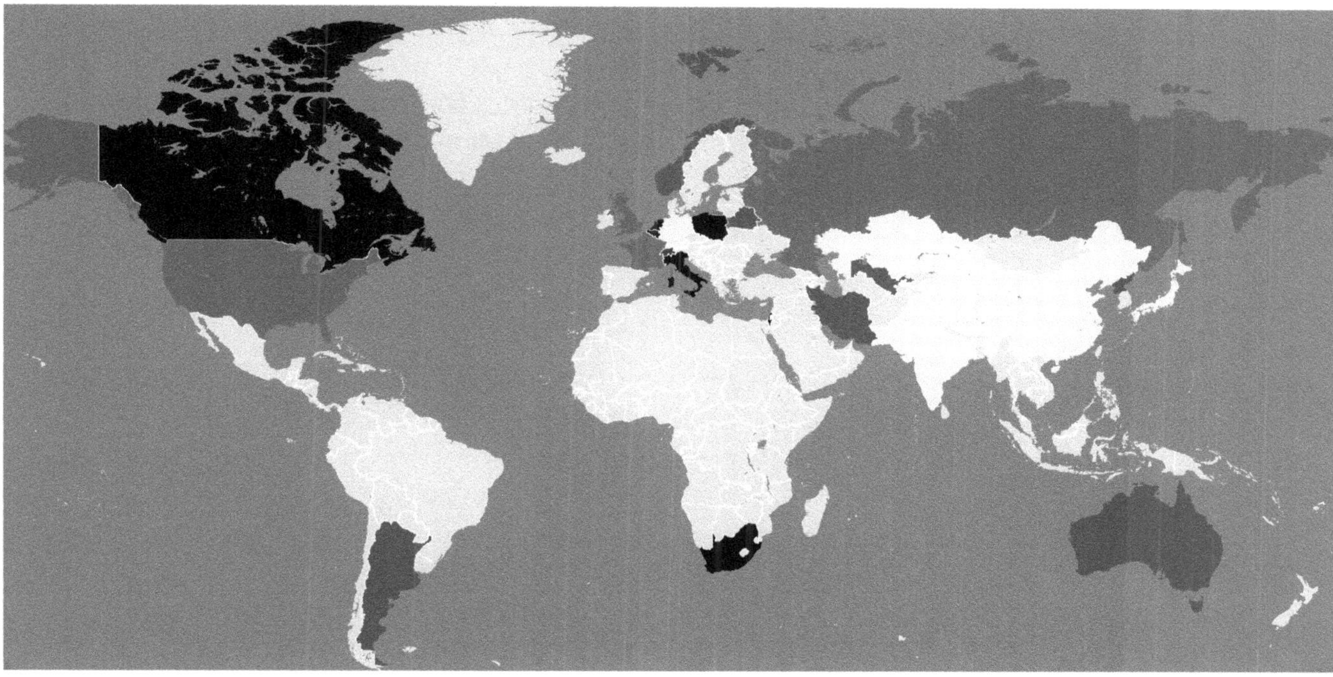

Less than 100 kg | 100 kg-1.99 tonnes | 2 tonnes-99.99 tonnes | 100 tonnes+

Nearly 2,000 metric tons of weapons-usable fissile material (highly enriched uranium or separated plutonium) are in 25 countries, potentially vulnerable to theft. The Nuclear Security Summit process has sought to reduce the risks of these stockpiles by establishing concrete commitments and cooperative measures for securing facilities and dangerous materials. However, the quantity of weapons-usable materials is increasing in India, Pakistan, the United Kingdom, and Japan.

- Acquiring sensitive technology under safeguards and then withdrawing from the NPT
- Developing the technical know-how for a nuclear-weapons program and an indigenous manufacturing capability

New entrants into the nuclear energy arena overall have been reluctant to disavow their "right" to sensitive technology and activities. In addition, these sensitive technologies that many admit pose proliferation risks are now even more easily acquired and assimilated because of advances in manufacturing techniques. North Korea and Iran are the two most notorious producers of fissile material, but several countries have expressed interest in acquiring these sensitive technologies for ostensibly legitimate civilian purposes, and many more desire to keep their options open.

Figure ES-2. Enrichment and Reprocessing Capabilities and Military Nuclear Programs

	ENRICHMENT	REPROCESSING	NON-PROLIFERATION TREATY MEMBER	MILITARY NUCLEAR PROGRAM
Argentina	X		X	PAST PROGRAM
Brazil	X		X	PAST PROGRAM
China	X	X	X	YES
France	X	X	X	YES
Germany	X		X	
India	X	X		YES
Iran	X		X	CURRENT STATUS UNCERTAIN
Israel	X	X		YES
Japan	X	X	X	
Netherlands	X		X	
North Korea	X	X		YES
Pakistan	X	X		YES
Russia	X	X	X	YES
United Kingdom	X	X	X	YES
United States	X		X	YES

Today, a handful of states host national uranium enrichment and spent fuel reprocessing facilities, which are the most sensitive portions of the nuclear fuel cycle. With the exception of Japan, Germany, and the Netherlands, these are all states that had or have nuclear weapons programs.

The world has been reluctant to confront this question: Do we really believe that we can live securely with a nuclear fuel cycle system that poses so few constraints on any state's ability to produce weapons-usable nuclear materials?

The Market

At the front end of the fuel cycle, global uranium enrichment capacity currently outpaces demand.[1] The combination of Japan's post-Fukushima nuclear reactor shutdowns, the German phase-out of nuclear power, and underutilized Russian enrichment capacity means that this will probably be the case for quite some time. And yet, despite this mismatch, new

1. Ux Consulting Company, "World Enrichment Nameplate Capacity," May 13, 2013, www.uxc.com/fuelcycle/Enrichment/uxc_EnrCapacityTable.aspx.

enrichment capacity being built today—in North Korea, Iran, and the United States—are national facilities and uneconomic.

At the back end, the amount of separated plutonium continues to grow internationally because it can't be consumed quickly enough. Decisions about reprocessing are often disconnected from use. This is evident in Japan's recent fuel cycle decisions and also in recent U.S. decisions about the mixed oxide (MOX) fuel program. As President Obama stated at Hankuk University in South Korea in 2012, "We simply can't go on accumulating huge amounts of the very material, like separated plutonium, that we're trying to keep away from terrorists."[2]

On long-term storage of spent fuel or disposal of nuclear waste, policy solutions remain elusive. Thirty-one countries (plus Taiwan) operate nuclear power plants, but no country has yet opened a final repository for waste. Geological disposal is the internationally accepted strategy, but political and technical difficulties so far have prevented the construction and operation of any commercial repositories. There is no functioning international market for disposal services at present, even though demand exists and the economics of a regional or multinational repository are encouraging. Meanwhile, spent fuel continues to accumulate in cooling pools with limited storage capacity. On-site dry cask storage is a longer-term but still temporary solution, and reprocessing and recycling still generate significant waste streams that require disposal. Continuing along this path is irresponsible and ultimately unsustainable. As the Blue Ribbon Commission on America's Nuclear Future observed: "This generation has a fundamental ethical obligation to avoid burdening future generations with the entire task of finding a safe permanent solution for managing hazardous nuclear materials they had no part in creating . . . while also preserving their energy options."[3]

There is no functioning international market for disposal services at present, even though demand exists and the economics of a regional or multinational repository are encouraging.

States seeking nuclear energy for the first time are increasingly asking for comprehensive fuel cycle services that encompass waste solutions. The nuclear fuel cycle market may look considerably different a few decades from now as the industry continues its consolidation

2. Barack Obama, "Remarks by President Obama at Hankuk University" (speech given at Hankuk University, Seoul, Republic of Korea, March 2012), http://www.whitehouse.gov/the-press-office/2012/03/26/remarks-president-obama-hankuk-university.

3. Blue Ribbon Commission on America's Nuclear Future, *Report to the Secretary of Energy* (Washington, DC: Blue Ribbon Commission on America's Nuclear Future, January 2012), http://cybercemetery.unt.edu/archive/brc/20120620220235/http://brc.gov/sites/default/files/documents/brc_finalreport_jan2012.pdf.

with multiple partnerships between private and government-backed organizations in all areas of the fuel cycle. There may be an opportunity to leverage current trends to shape fuel cycle services in a way that is acceptable to industry and better serves nuclear security and nonproliferation goals.

Why Does This System Persist?

Experts have studied alternatives to national control of fuel cycle facilities for more than 60 years, developing proposals that either did too little or did too much. None of the proposals focused on the underlying issues: strongly held positions on sovereign rights, and decision-making that has become inconsistent with how the world addresses many other security concerns in which there are shared responsibilities and is exacerbated by the uneven distribution of global capabilities.

The nonproliferation regime has relied upon voluntary supplier controls for the past 40 years, made easier by the concentration of 80 percent or 90 percent of supply in the hands of four or five suppliers. The nonsuppliers, shut out of a market, have become deeply mistrustful of even the mildest attempts to shore up the perceived status quo, including supply assurances on the front end (e.g., fuel banks).

What's more, some states object in principle to potential restrictions on freedom of decisionmaking with regard to nuclear energy and fuel cycle, including some who have no near-term plans to develop or introduce such technology and would therefore not be affected by changes to the system. In addition, some non-nuclear-weapon states would like to tie further restrictions to progress by the nuclear-weapon states on nuclear weapons reductions and disarmament. Other states—including some with nuclear weapons and existing or planned nuclear facilities—choose political and economic caution over changes to the status quo.

Any alternative, therefore, has to be politically acceptable across a wide spectrum of states. Fundamentally, however, all parties—industry, government, and publics—share the goal of enhanced access to fuel cycle products and services. The question is whether proposals offer enough incentives for states to accept fuel cycle restrictions or offer enough controls so that no fuel cycle restrictions are needed.

Objectives for Production, Use, and Disposal of Fissile Material

The spread of technologies to produce nuclear fuel for reactors—uranium enrichment and spent fuel reprocessing—is the most challenging of the dangers presented by the existing system. The most basic premise for rethinking the fuel cycle is that the risk of misuse of such facilities can be reduced by removing these facilities from national control. However, *simply changing control and/or ownership structures is not enough to prevent, deter, or better detect proliferation.* It is necessary to manage new activities and identify a path for transitioning existing fuel cycle facilities in a way that promotes political, economic, institutional, and technical sustainability. To further address these risks, uranium enrichment should be in balance with global demand, new facilities should be under diversified partnerships, HEU should no longer be produced for civil purposes, and plutonium should only be separated from spent fuel when use (in peaceful fuel) is imminent.

The good news is that civilian and military HEU production worldwide is already voluntarily limited, but it should be eliminated for civilian purposes under this new approach. Any new enrichment facilities need to meet specific criteria (see the following section on best practices) or the state would be subject to a general ban on nuclear-related commerce. New reprocessing facilities would also need to meet specific criteria, including multinational ownership/operation, willingness to accept spent fuel from others, and operation scaled to consumption of separated plutonium. A fundamental objective is to have no net increase in the amount of nuclear-weapons-usable material. For research and development, the long-term goal should be a phaseout of technologies that result in separated plutonium being produced and stockpiled.

Efforts now focusing on minimizing civilian HEU should set their sights on eliminating HEU. Some of the existing stockpiles can be earmarked for use until the full technical transition is complete and the rest should be downblended to LEU. The tons of separated plutonium that exist around the world need to be disposed of or consumed, either by mixing it with waste and burying it or burning plutonium as fuel in advanced reactors. In all cases, the objective here would be to have no net increase in nuclear-weapons-usable material, and to use existing material before production of new material where there are conversion and consumption paths available.

The good news is that civilian and military HEU production worldwide is already voluntarily limited, but it should be eliminated for civilian purposes under this new approach.

For disposal, a primary objective is to create financial and other incentives for new storage facilities and repositories. Therefore, a new approach should offer access to repositories to countries that do not pursue national fuel cycle facilities. Options for managing spent fuel could include fuel leasing/take-back, fuel cycle parks, regional/international repositories, and interim storage. Diversified partnerships for repositories, (i.e., reserved storage space for partners and consideration of diversified or regional approaches to new

Figure ES-3. Likely Outcomes of the Current Path and a New Approach

	CURRENT PATH		NEW APPROACH
PRODUCTION (ENRICHMENT)	New national enrichment facilities with questionable economic justification	>	Enrichment in balance with global demand; new facilities under diversified partnerships; no HEU use for civil purposes
PRODUCTION (REPROCESSING)	New separation and material accumulation	>	Separation only when linked to use
USE	Slow progress on programs to consume or dispose of separated material	>	New technologies and programs to consume, phased to match separation timelines after addressing existing stockpiles
STORAGE/ DISPOSAL	Few new repositories, not enough capacity for current SNF inventories	>	Long-term storage and repositories serving multiple clients/countries

interim storage) should be careful not to undermine any existing successful national programs for waste disposal.

Best Practices for Shaping a New Nuclear Architecture

The above-mentioned objectives for production, use, and disposal of fissile material/nuclear waste need to be carried out in such a way that improves the political, economic, institutional, and technical sustainability of nuclear energy. The new approach would need to incorporate the criteria below:

- Equal access and shared benefits
- Shared costs and burdens
- Early detection

- Minimizing weapons-usable material
- Market-driven expansion
- Incentives and consequences

Applying these "best practices" to fuel cycle activities will require changes to current arrangements across diverse industrial sectors and diverse domestic and technical environments.

EQUAL ACCESS AND SHARED BENEFITS

States with fuel cycle facilities should expand opportunities for those without to invest in commercial ventures and share in the benefits, without increasing security risks.

Opening up access and benefits to nontechnology holders will require alternative operating and financial arrangements for new and existing facilities. Diversified partnerships for ownership and operations of new facilities and retroactive phase-in for existing facilities could include financial investment, equity stake, and/or rights to output and product, among other things. Rules and guidance should be nondiscriminatory and equitably applied with respect to access and benefits. These partnerships could be particularly important for creating regional storage or disposal sites for spent nuclear fuel. With respect to reprocessing, the significant costs of building facilities provide an incentive to states to seek outside investment, and there may be economic and political incentives on both supplier and recipient sides for widening ownership.

SHARED COSTS AND BURDENS

All facilities would follow established and evolving best practices for security, safety, and safeguards for all materials and facilities in use or in transit. Government and industry would contribute to a consumption/sustainability tax to pay for additional safeguards and any other burdens on the International Atomic Energy Agency (IAEA).

Current technology holders should accept the same restrictions as new entrants so that the changes to the system would not be disproportionately felt by nontechnology holders. Applying IAEA safeguards to all existing enrichment and reprocessing plants and the conclusion of Additional Protocols may be possible within 10 years. In addition, all facilities utilizing HEU or Pu (such as MOX fuel fabrication plants) will need to be safeguarded in order to provide early warning of misuse. Needless to say, all operating enrichment and reprocessing facilities should implement internationally accepted standards and good practices for security and safety (including IAEA INFCIRC-225 and other relevant guidance). In all cases, the establishment of an independent regulator is an essential element. In some cases, the additional financial burden of activities required under this approach may require a sustainability "tax" on facility operators—perhaps scaled to the necessary effort.

EARLY DETECTION

The system should maximize the likelihood of early detection of noncompliance, including through enhancing transparency, to improve the ability of the international community to intervene in a timely fashion.

As practiced now, access to sensitive technology is controlled ostensibly for nonproliferation reasons, but guidelines are set commercially for proprietary reasons. Moreover, there is no uniformity across centrifuge producers. Experts need to improve consistency on how sensitive technology and processes—including manufacture, installation, and maintenance—are controlled through limiting the ability to copy and build components, constraining modification of operating facilities, and limiting advance planning time. These measures will help ensure that detection is early, credible, and unambiguous, facilitating intervention in case of breakout.

MINIMIZING MATERIAL

The system should yield the least possible amount of weapons-usable material to reduce proliferation and security risks.

States must make political commitments and regulatory assurances that there will be no HEU production, no new facilities licensed for HEU production or use and no civil HEU-fueled facilities, and no plutonium separation unless there is a concurrently available path for consumption. Technical and commercial partnerships to explore ways in which fast reactor research can be oriented toward minimizing HEU use and Pu production should be expanded and politically supported. Countries also need to explore whether reactors, reprocessing, and fuel fabrication activities should be collocated in nuclear "islands," and whether these islands should be under multinational control.

MARKET-BASED EXPANSION

No new facilities capable of producing weapons-usable material should be constructed unless there is unmet commercial demand, to limit excess, underutilized material production capacity.

Market-based expansion is already the foundation of commercial activities for the front end of the fuel cycle, but countries need to agree that any new capacity should meet a market test and only when there is a commercial driver should capacity be expanded. Regarding spent fuel management and disposal, there is an unmet market need for national, regional, or international joint approaches. Through cooperation, this market demand can be met but requires high level political decisions based on a national or international security assessment.

INCENTIVES AND CONSEQUENCES

Governments and industry should benefit more from adhering to their commitments than from breaking them.

Without a treaty, incentives and consequences for implementing best practices are essential. Three potential avenues for incorporating incentives and consequences include commercial (such as contractual obligations and codes of conduct), national (such as nuclear cooperation agreements), and international (such as treaties or other agreements). Incentives could include prior consent for certain activities. Eventually, in a fully realized system, consent rights could become unnecessary and obsolete. Consequences for not implementing best practices could include targeted limits on uranium supply to the country (or even spent nuclear fuel); conversion services; fuel fabrication services; ability to purchase or use the product of fuel cycle facilities; ability to send high-level waste or spent fuel to a repository; or commercial interactions by third parties with any nuclear entity in the country. Taken together, these limits would constitute a general ban on nuclear-related international commerce for the country in question. The judgment of whether activities meet new best practices could be drawn by individual states or commercial entities. Or, it could be delegated to the IAEA. This evaluation could begin immediately for new facilities and be tied to a defined transition plan for existing facilities (based on national and corporate policies).

Altogether, this new approach would counter the current de facto monopoly of the fuel cycle owner-states and increase access to the peaceful benefits of nuclear energy.

Altogether, this new approach would counter the current de facto monopoly of the fuel cycle owner-states and increase access to the peaceful benefits of nuclear energy through a layered system that includes the active participation of states in facilities and fuel assurances. Current technology holders would have to accept new partners and regulatory structures, so the changes to the system would not be disproportionately felt by nontechnology holders. In order to make the decision to support a change easier for a state-owned entity or private company, new approaches to the operation of fuel cycle facilities should also address concerns about energy security and nondiscrimination.

Building Support for the Action Plan

The most critical step is to help build support for a best practices approach to the nuclear fuel cycle. Discussions on the unsustainability of the current path should begin on a Track 1.5 basis, continuing the work of several organizations in the NGO community of the last few years. More targeted discussions among government, industry, and stakeholders (including international organizations and the policy community) could focus on the following:

- Black-box approaches to sensitive fuel cycle facilities, including work to rationalize approaches designed to protect industry secrets and those designed to prevent proliferation of sensitive technologies
- Fuel cycle decisionmaking in countries pursuing advanced nuclear fuel cycles (in contrast to technical discussions), particularly in Russia, China, and India

- Security and safeguards requirements for older spent nuclear fuel, particularly low burn-up fuel
- Desirability (technical and political) of collocation of reactors, reprocessing, and fuel fabrication in nuclear islands, potentially under multinational control
- The applicability of lessons of the European spent fuel management experience in other regions, particularly Asia and the Middle East
- Perceptions (by governments and industry) on security of supply, and on methodologies for determining market need for fuel cycle services.

Existing fora, such as the International Forum for Nuclear Energy Cooperation (IFNEC), the Nuclear Suppliers Group (NSG), the Asia-Pacific Safeguards Network, or the IAEA, could address these topics, but it will be important to engage industry fully in this process. To this end, a standing joint industry-policy community forum for discussion of security and proliferation concerns related to the nuclear fuel cycle should be created. Such a forum could help draft, for example, fuel cycle industry principles of action that would support a new best practices approach.

It is likely that a "friends of sustainable nuclear energy" group that is widely representative of the governments and industries of supplier and recipient countries will be necessary to help cement a more sustainable approach to the nuclear fuel cycle in the future.

Figure ES-4. Action Plan for Implementation

BEST PRACTICE	STEPS	WHO	HOW	TIME FRAME
EQUAL ACCESS AND SHARED BENEFITS	Diversified partnership for enrichment, reprocessing facilities	Companies (private industry, state-owned, and hybrid)	Examine existing models, define appropriate diversification models	10 years for existing, from beginning for new
	Defined waste path	States, suppliers, and operators	Continue international cooperation, include in supply contracts	10 years for interim approaches, 30 years for long-term disposal
SHARED COSTS AND BURDENS	Safeguards: Additional Protocol and international safeguards on <u>all</u> civilian enrichment and reprocessing (E/R) plants	States, IAEA	Conclude agreements, implement safeguards, amend voluntary offer safeguards agreements for states without comprehensive safeguards	5 years for conclusion of APs, 10 years for implementation of safeguards at E/R plants
	Implementation of INFCIRC-225 (most recent revision)	States	Secure public commitment, embed in legal, regulatory systems	3 years
	Independent regulator	States	Ensure through domestic legal and institutional authorities	5 years
	Sustainability tax	States (via regulators) and IAEA	Define who will need resources for obligations (IAEA, etc.), negotiate arrangements	10 years, and paired to above-stated commitments so that there is no gap
EARLY DETECTION	Black-box and operational separation	New enrichers, technology producers, and operators	Engage in expert-level dialogue, determine applicability to known technology, define appropriate separation of roles	From beginning for new facilities

Figure ES-4. (cont.)

MINIMIZING WEAPONS-USABLE MATERIAL	No production or use of HEU	States and regulators	Require policy commitments, voluntary reporting, technical R&D for conversion, international cooperative programs for removals	Immediately following policy commitments, phased in as technical approaches are validated
	Drawdown of HEU and Pu stockpiles	Operators, states, and regulators	Secure policy commitments, licensing for utilization of Pu, downblending of HEU	Immediately following policy commitments and technical assessment of facilities; 20–30 years for drawdown of excess inventory
MARKET-DRIVEN EXPANSION	None-already driving commercial activity for enrichment	Companies	Continue limiting new capacity until there is a market need	Current and ongoing
	Development of regional and international options to meet market demand for the back end	States, regulators, and companies	Create consortia, explore technical and siting options, define collective needs	Beginning immediately for option development; 10–30 years for implementation
INCENTIVES AND CONSE-QUENCES	Requiring prior consent for processing of supplier's material and/or equipment	States and suppliers	Include in nuclear cooperation agreements and standard supply contracts; provide opportunities for material uses when consistent with the best practices	Beginning immediately for new agreements and phased in over 15 years for existing as they are renewed

INTRODUCTION

THE NUCLEAR NON-PROLIFERATION TREATY (NPT) DECLARES WHAT MANY INTERPRET AS AN UNRESTRICTED, SOVEREIGN RIGHT TO NUCLEAR TECHNOLOGY FOR PEACEFUL PURPOSES. When the treaty was negotiated in 1968, nuclear fuel cycle technology was believed to be beyond the reach of all but a few countries and could be protected and contained. This proved false.

Interest in managing the nuclear fuel cycle has risen and fallen with the fortunes of the civilian nuclear energy industry. In the past decade, a resurgence of enthusiasm for nuclear power has rekindled interest in efforts to manage the fuel cycle. Although some countries may be reevaluating their nuclear energy plans in the wake of the 2011 accident at the Fukushima Daiichi nuclear power plants in Japan, current proliferation crises in North Korea and Iran raise this question: Does the current approach to the fuel cycle—leaving uranium enrichment and spent fuel reprocessing capabilities in the hands of national governments—create too great a proliferation and security risk? There is a need for new approaches to the nuclear fuel cycle that can help mitigate proliferation risks, make nuclear energy safer and more sustainable, and help improve nuclear governance. To that end, the Nuclear Threat Initiative and the Center for Strategic and International Studies launched the New Approaches to the Fuel Cycle project in early 2011.

Experts have studied alternatives to national control of fuel cycle facilities for more than 60 years and developed a range of proposals that addressed technical concerns but did not get traction among industry actors and policymakers. This is at least in part because there is no consensus on the nature of the problem with current arrangements, or even on whether there is a problem at all. Proposals have more often focused on the front end of the fuel cycle—uranium enrichment—than on the challenges of managing spent fuel. None of the proposals focused on the underlying issue: a perception of sovereign rights and decisionmaking that is inconsistent with how the world addresses many other security concerns and is exacerbated by uneven distribution of global capabilities. Meanwhile, the promise of proliferation-resistant technologies remains unrealized.

This report aims to build consensus on common goals, recognize practical challenges, and engage a broad spectrum of actors who can influence policymaking regarding the nuclear fuel cycle. It outlines a vision for an integrated approach to nuclear supply and demand, drawing on international expertise from industry, government, and nongovernmental organizations. The project, which hosted multiple workshops and smaller breakout groups to vet ideas, sought specifically to identify practical solutions that could be adopted in phases. The project used "best practices" as an integrating theme to describe a sustainable approach across elements of the fuel cycle and across different sectors. It also tackled one of the toughest issues—spent nuclear fuel and high-level waste—to see if solutions there might offer incentives to states on the front end of the nuclear fuel cycle.

Incentives are important because efforts to improve the robustness of the nonproliferation regime face two challenges: inertia and concern that change may bring additional burdens and restrictions. Yet all parties—whether they represent industry, government, and the general public—share the goal of enhanced access to fuel cycle products and services. This report outlines best practices to achieve both goals: improvements to legitimate access, and reducing proliferation and security risks.

1. THE NEED FOR A NEW APPROACH: CURRENT PROLIFERATION AND SECURITY RISKS

THE PROLIFERATION AND SECURITY RISKS INHERENT IN THE CURRENT NUCLEAR FUEL CYCLE ARE NOT JUST TECHNICAL IN NATURE, BUT HAVE POLITICAL AND ORGANIZATIONAL FACETS AS WELL. At the most basic technical level, enrichment and reprocessing facilities generate fissile materials that can be used for either civilian or military purposes (See Figures 1 and 2). Although most material produced for civilian purposes is not weapons-grade, this is not an insurmountable obstacle to proliferation. Once a state has produced enough LEU enriched to roughly 4 percent in U-235, it has already completed about three-quarters of the enrichment effort required to produce weapons-grade uranium.[1]

Reprocessing plants and associated storage and fuel fabrication facilities pose particular proliferation and security risks because they usually have substantial stocks of separated plutonium. Although probably not the material of choice for proliferators, reactor-grade plutonium has been and could be used in nuclear explosive devices.

Moreover, fully closing the fuel cycle with fast neutron reactors dependent on plutonium-based fuels would add to proliferation risks. Such an approach, if widely adopted, would result in hundreds of tonnes of separated plutonium being put into commercial circulation, increasing the possibility of diversion for illicit weapons programs. If those fast reactors are configured to breed more plutonium than they burn, the risk is greater. Critics argue there is no economic justification for a closed fuel cycle because the once-through cycle is cheaper and uranium is abundant. Efforts to reduce the proliferation risks of closed fuel cycles, through either avoiding the use of HEU or Pu in the start-up fuel or by designing "burner" reactors that can use fuel that has been conditioned rather than reprocessed (i.e., where the plutonium is not fully separated from transuranics), are in development but may take decades to yield results. In the meantime, promotion of a closed fuel cycle continues to exert pressures on an already-overtaxed international safeguards system.

---

1. Sidney Drell et al., *Verification of Dismantlement of Nuclear Warheads and Controls on Nuclear Materials* (McLean, VA: MITRE Corporation, January 1993), 59, http://fas.org/irp/agency/dod/jason/dismantle.pdf.

Figure 1. National Stocks of Separated Plutonium

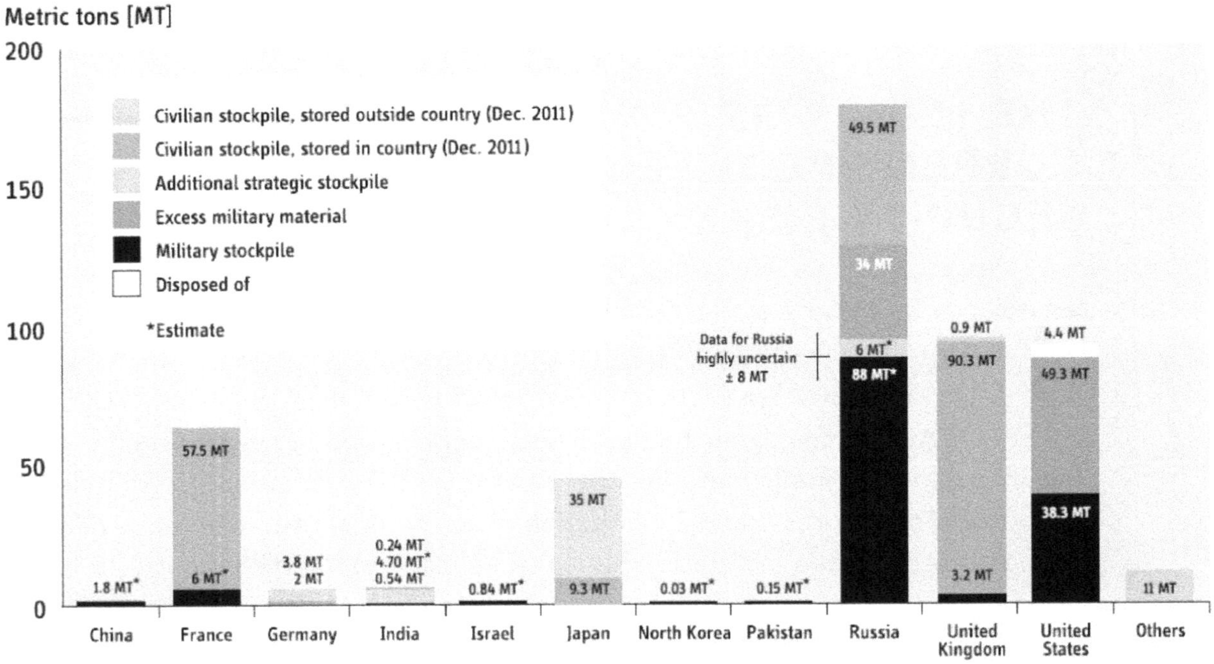

Source: International Panel on Fissile Materials (IPFM), Global Fissile Material Report 2013 (Princeton, NJ: IPFM, 2013), 18, http://fissilematerials.org/library/gfmr13.pdf.

Beyond the duality of the material, the production processes pose other technical risks. First, these facilities handle material in bulk (versus discrete items that can be counted) and often process so much material that error margins in accounting could hide diversion of several bombs' worth of material. Second, newer technologies for enrichment might make it easier for diversion (e.g., changing valve switches instead of piping, or laser tuning). What's more, a national program that includes manufacturing of enrichment components could provide the opportunity to establish a parallel clandestine enrichment program.

Several states already host national enrichment and reprocessing (E/R) facilities,[2] which are the most sensitive elements of the nuclear fuel cycle, and new entrants into the nuclear energy arena are reluctant to forswear development in this area.

Countries such as Russia, China, and India are looking for assistance, either financial or technical or both, to further develop and expand their own fuel cycle capabilities. Russia, for example, has sought technical assistance with reprocessing from France (though

2. This report uses the terms "reprocessing" and "spent nuclear fuel," recognizing that advocates of the closed fuel cycle prefer the terms "recycling" and "used nuclear fuel." The term "recycling" is used in this report to refer to the reuse in reactors of nuclear material recovered from irradiated fuel. We have chosen not to use the term "used nuclear fuel" because it is not as widely accepted currently as the term "spent nuclear fuel"; this is not meant to imply any statement about the usability of the irradiated fuel.

unsuccessfully) and suggested that it would welcome outside investment in fuel cycle facility capacity expansions, and China has purchased a centrifuge enrichment plant from Russia, albeit on a black-box basis. Countries interested in developing nuclear energy for the first time (Saudi Arabia and Jordan, for example), or that are expanding existing programs (South Korea), are reluctant to give up any right to domestic enrichment and reprocessing operations.

India and Pakistan present a special set of challenges because both are ramping up plutonium production. In Pakistan's case, this expansion appears to be solely related to increased weapons production. India now has four reprocessing plants, only one of which is subject to intermittent IAEA safeguards, and more capacity expansions are planned. One of these plants is earmarked for military use while the others are described as dual-use, meaning they could be used to produce plutonium for India's planned fast breeder program or for nuclear weapons. India is also rapidly expanding enrichment capacity for its military program.[3]

Figure 2. National Stocks of Highly Enriched Uranium (HEU)

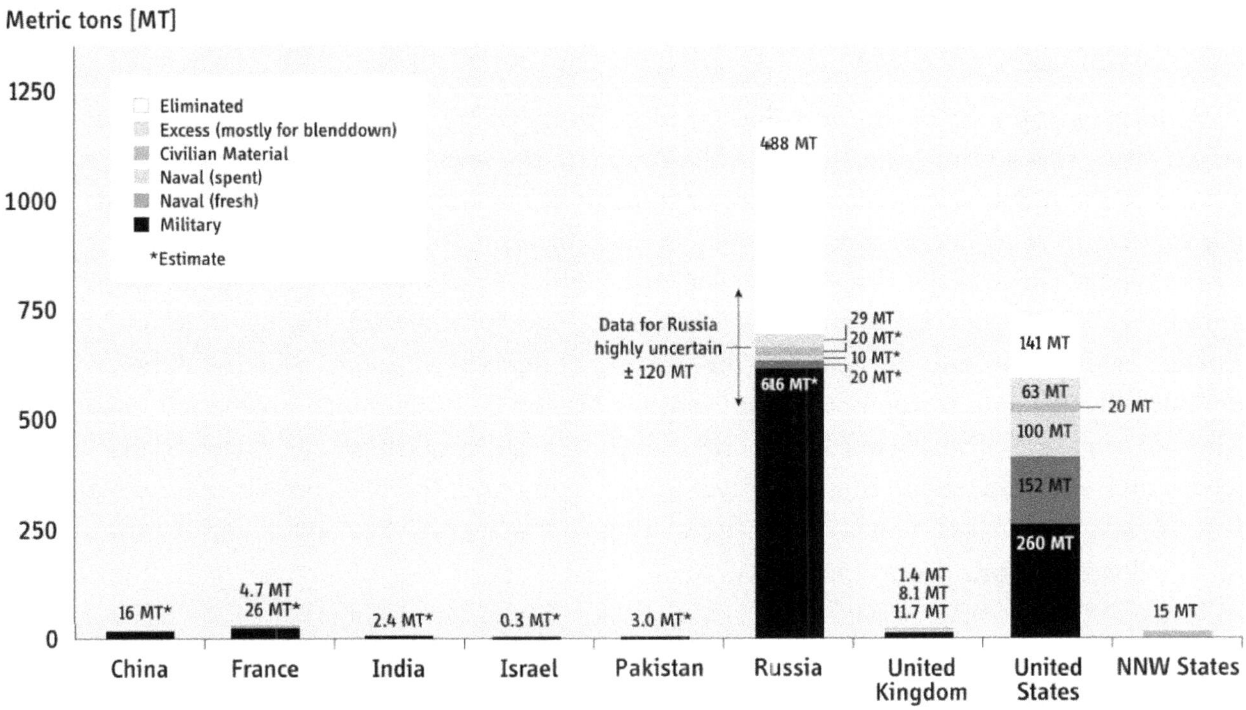

Source: IPFM, Global Fissile Material Report 2013, 11.

3. Douglas Busvine, "India nuke enrichment plant expansion operational in 2015—IHS," Reuters, June 20, 2014, http://in.reuters.com/article/2014/06/20/india-nuclear-idINKBN0EV0JR20140620.

Fast breeder reactors, such as Argonne National Laboratory's EBRII shown here, have been developed by six countries (five nuclear weapon states and Japan) beginning in the 1950s, but few have been deployed commercially because they have not proven to be economically viable.

These "virtual" weapons capabilities strain the international safeguards system and complicate detection of diversion to military uses. However, detection does not equal constraint, and activity carried out at declared facilities can move states further along the learning curve toward a weapons capability and shorten the time to a bomb.

During negotiation of the Non-Proliferation Treaty (NPT), diplomats discussed where exactly they should draw the line on proliferation. In the end, the NPT restrictions were limited to the manufacture, acquisition, or transfer of nuclear weapons or their control or the seeking or receiving of any assistance in manufacturing nuclear weapons. Some diplomats questioned whether drawing the line that high up in a long chain of decisions was adequate. The Soviets, for example, wanted to include "preparation for manufacture" as well as, at the other end of the spectrum, testing of nuclear weapons. Swedish Ambassador Alva Myrdal, in a statement to the Eighteen Nation Disarmament Conference, asked whether a middle ground could be found that prohibited acquisition and enjoined states to refrain from preparation of nuclear weapons.[4] In the end, the NPT prohibition on manufacture, acquisition, and control permitted a range of activities that could help contribute to a nuclear-weapons program.

How much does any of this matter? The concern is that a state within the treaty could slowly and stealthily acquire capabilities to make nuclear weapons without actually violating the NPT. These dangerous gaps are unlikely to be fixed in the treaty itself. Instead, suppliers have implemented voluntary restraints since the mid-1970s.

---

4. Alva Myrdal, "Statement by the Swedish Representative (Myrdal) to the Eighteen Nation Disarmament Committee: Nonproliferation of Nuclear Weapons, February 24, 1966," in *Documents on Disarmament*, ed. Robert W. Lambert (Washington, DC: U.S. Arms Control and Disarmament Agency, September 1967), 56, http://www.un.org/disarmament/publications/documents_on_disarmament/1966/DoD_1966.pdf.

Systemic weaknesses in the International Atomic Energy Agency (IAEA) safeguards form another layer of technical risks. The primary objective of IAEA safeguards, per Article III.1 of the NPT, is to prevent "diversion of nuclear energy from peaceful uses to nuclear weapons or other nuclear explosive devices." The IAEA has set goals for timely detection to enable the international community to intervene before a state attempting a diversion is able to produce nuclear weapons. But this system has two problems. First, at bulk-handling facilities such as uranium enrichment and spent fuel reprocessing plants, the quantities of material that statistically fall into the norms of acceptable "material unaccounted for" exceed the levels that the IAEA has set for "significant quantities." Verification of nondiversion must be based on additional information provided by containment and surveillance measures (seals and cameras); material accountancy alone does not reduce uncertainties to an acceptable level. This problem is multiplied in facilities with large throughputs. Second, although detection of anomalies may be timely, delays in reporting from the IAEA Secretariat and in action by the Board of Governors can reduce warning time dangerously.

Political and institutional arrangements pose additional challenges to nonproliferation efforts. The IAEA Board of Governors and/or the IAEA Secretariat can conclude that a state is in noncompliance with its safeguards agreement. In the case of the Board, politics clearly can enter into the picture, although the

"Rights" or Risks under the Nuclear Non-Proliferation Treaty?

The following steps, although in full compliance with the NPT, could aid a clandestine nuclear-weapons program.

- Produce and stockpile highly enriched uranium (>20 percent U-235) for any purpose (e.g., research reactor fuel, small modular reactor fuel, naval propulsion fuel, and medical isotopes).
- Withdraw highly enriched uranium from safeguards by declaring certain components of a program to be intended for nonexplosive military purposes.
- Accumulate stocks of separated plutonium. (Nuclear Suppliers Group members adhere to some restrictions, but there is no legal barrier to the transfer of separated plutonium.)
- Accumulate stocks of highly enriched uranium, low-enriched uranium, or plutonium.
- Acquire sensitive technology under safeguards, then withdraw from the NPT and retain control of the technology.
- Develop the technical know-how for a nuclear-weapons program and an indigenous manufacturing capability.

Board is not bound absolutely by consensus rules in finding noncompliance (as was the case for Iran). In the case of the Secretariat, systemic weaknesses in the safeguards system prior to 1991 made determinations of noncompliance difficult, in particular because efforts focused on correctness and not completeness of declarations. Although the IAEA has worked hard to overcome these challenges through the adoption of the Additional Protocol and other steps, the fact that the Board found Iran in noncompliance with its safeguards agreement, despite inspectors not reporting noncompliance from the field, is one indicator that these decisions can be difficult even when evidence is overwhelming. Lastly, the United Nations Security Council has limited ability to enforce resolutions and affect behavior even when the IAEA does report noncompliance.

The Fuel Cycle and the Current Nonproliferation Regime

An international system, underpinned by the NPT, bilateral nuclear cooperation treaties, IAEA inspections, and international export controls, provides some reassurance to governments and the public that the use of nuclear energy across the globe can take place with a reduced risk of diversion for military purposes than might otherwise be the case.

The IAEA safeguards system is a foundation of the nonproliferation regime. Since the early 1990s, the IAEA has strengthened its safeguards system to look for indicators of undeclared activities as well as the nondiversion of declared material. Under the Additional Protocol (INFCIRC/540), the IAEA added technical tools, expanded its access, and enhanced reporting requirements.

The IAEA has been involved for decades in helping shape technical fuel cycle developments among member states. Some examples include its role in convening the International Nuclear Fuel Cycle Evaluation study, activities under the International Project on Innovative Nuclear Reactors and Fuel Cycles, and coordination with the Generation IV Forum on advanced reactors. Although the IAEA can be a key promoter of improved proliferation-resistance, it cannot actually shape national decisions; instead, it responds to the wishes of its member states. In 2003, however, then Director General Mohamed ElBaradei strongly advocated efforts to multilateralize sensitive fuel cycle facilities in an article in *The Economist*. This was followed by a 2005 study (described later in this report) on multilateral approaches to the nuclear fuel cycle.

A second element of the nonproliferation regime for mitigating risks from sensitive nuclear technologies is the system of export controls. These are implemented on the basis of national laws but also within the Nuclear Suppliers Group (NSG), within which suppliers harmonize their export control guidelines.

The International Atomic Energy Agency (IAEA) based in Vienna, Austria is a key element of the global nuclear nonproliferation regime, including implementing safeguards on nuclear materials and shaping nuclear fuel cycle technical developments.

Before the IAEA was established, and in the decades before the NPT entered into force, countries concluded bilateral agreements that included conditions for using nuclear material and equipment. When IAEA safeguards largely replaced bilateral safeguards, some bilateral restrictions were still considered necessary, and this continued after the NPT entered into force because the treaty contained so few restrictions on material transfers and processing. With the objective of avoiding situations where material is misused for military purposes, suppliers therefore restricted where and how material can be stored and/or transferred to third parties and whether material can be subsequently enriched or reprocessed.

These restrictions are not uniformly applied, however. Even the United States, which has the most restrictive regulations based on the 1978 Nuclear Non-Proliferation Act that amended the Atomic Energy Act of 1954, makes exceptions for some countries. For transfers, the United States has granted consent on a shipment-by-shipment basis for transfer of U.S.-obligated spent nuclear fuel (SNF) to the European Atomic Energy Community (Euratom) for reprocessing, retaining prior consent rights on the further disposition of recovered plutonium. The United States also gave advance consent for Japan to retransfer natural uranium and LEU to other countries, but not for enrichment at or above 20 percent. In the case of consent for enrichment and reprocessing, the U.S. policy has been tailored according to nonproliferation criteria that a recipient country must meet. The United States has granted advance consent (a one-time, blanket approval) to a handful of countries, including Japan and Euratom. Other countries take varied approaches.[5]

The Nuclear Suppliers Group, formed in the mid-1970s, slowly expanded out from the "big ticket" items to a longer list of items, including dual-use items (i.e., equipment that could be used in sensitive nuclear fuel cycle facilities but had other non-nuclear commercial applications). With respect to fuel cycle capabilities, the NSG until recently followed a policy of restraint. The key provision in NSG Guidelines (INFCIRC/254/Rev 1./Part 1) is a commitment to "restraint in

5. Fred McGoldrick, *Nuclear Trade Controls: Minding the Gap* (Washington, DC: CSIS, January 2013), 34, http://csis.org/files/publication/130122_McGoldrick_NuclearTradeControls_Web.pdf.

the transfer of sensitive facilities, technology, and material" (i.e., enrichment and reprocessing facilities, equipment, and technology), and, more generally, a nonproliferation principle, which states that "suppliers should authorize transfer of items . . . only when they are satisfied that the transfers would not contribute to the proliferation of nuclear weapons or other nuclear explosive devices or be diverted to acts of nuclear terrorism."

From 2004 to 2011, NSG member states discussed how to strengthen criteria regarding transfers of enrichment and reprocessing technology. The final paragraphs in the NSG guidelines added details but broke very little new ground. States petitioning for E/R facilities, equipment, or technology are eligible for transfers if they are members in full compliance with the NPT; have not been identified as being in breach of their safeguards obligations; adhere to NSG guidelines and United Nations Security Council Resolution 1540; have concluded an intergovernmental agreement on nonexplosive uses, safeguards in perpetuity, and retransfers; and implement physical protection and safety procedures according to international standards. Paragraph 7 of the guidelines details a set of arrangements under which such facilities, equipment, and technology may be transferred, including no enrichment above 20 percent, protection of technology, etc.

In the protracted discussions, NSG members considered other criteria with regard to transfers, including whether the transfer would have a negative impact on the stability and security of the region, whether the recipient has a credible and coherent rationale for pursuing enrichment and/or reprocessing capabilities in support of civil nuclear power programs, whether the recipient is party to an agreement not to pursue enrichment or reprocessing, and whether it is likely that a transfer would provoke countries in a region to seek or acquire sensitive nuclear technologies. These criteria were ultimately not included in the new language of INFCIRC/254.

Although NSG guidelines are not legally binding, NSG members largely have not transferred E/R to nontechnology-holding states since adoption of the guidelines some 40 years ago. However, member states at times have ignored the guidelines in the interest of export revenues and/or geopolitical considerations, and an underground network of nuclear suppliers has been able to ferret out information and supplies to countries shunned by NSG supplier states. In the future, this approach risks even greater challenges as developing countries grow more confident in their ability to design and produce "homegrown" nuclear fuel cycle facilities, based on technical information that can be obtained without breaking any national or international laws.

The Politics of Tightening Restrictions

Any discussion of limiting national enrichment and reprocessing facilities consistently runs up against long-standing political differences. Developing countries view this debate as yet another attempt by developed countries to curtail their growth and deny equal access to advanced technologies. This sentiment is also driving some states' refusal to sign or implement the Additional Protocol. Deep mistrust among non-nuclear-weapon states is linked to the perceived lack of progress by the five nuclear-weapon states to make progress on nuclear disarmament, in effect, failing to hold up their end of the "grand bargain."

Some countries claim NSG guidelines to restrict technology and equipment transfers violate the obligations under NPT Article IV to "undertake to facilitate and have the right to participate in the fullest possible exchange of equipment, materials, and scientific and technological information for the peaceful uses of nuclear energy." Even within the NSG, members have been hesitant to turn restrictions on themselves because nontechnology holders did not want to rule out future capabilities. Keeping options open apparently applies equally to NSG members and the nonaligned.

The current trajectory of fuel cycle policy and pursuits puts us on a path of increasing proliferation and security risk over the long term, as illustrated in Figure 3.

With respect to enrichment, there is little commercial need for increased capacity for the foreseeable future. Global enrichment capacity currently totals roughly 55 million separative

Figure 3. Likely Outcomes of the Current Path and a New Approach

	CURRENT PATH		NEW APPROACH
PRODUCTION (ENRICHMENT)	New national enrichment facilities with questionable economic justification	>	Enrichment in balance with global demand; new facilities under diversified partnerships; no HEU use for civil purposes
PRODUCTION (REPROCESSING)	New separation and material accumulation	>	Separation only when linked to use
USE	Slow progress on programs to consume or dispose of separated material	>	New technologies and programs to consume, phased to match separation timelines after addressing existing stockpiles
STORAGE/ DISPOSAL	Few new repositories, not enough capacity for current SNF inventories	>	Long-term storage and repositories serving multiple clients/countries

work units (SWU), while global demand totals roughly 51 million SWU.[6] Japan's post-Fukushima nuclear reactor shutdowns, the German phaseout of nuclear power, and underutilized Russian enrichment capacity mean that this will probably be the case for quite some time. Enrichment facilities currently being built (in North Korea, Iran, and the United States) are all nationally owned and uneconomic.

With respect to reprocessing, the amount of separated plutonium continues to grow internationally because it can't be consumed quickly enough. Decisions about reprocessing are often disconnected from use. This is evident in Japan's recent fuel cycle decisions and as well as in recent U.S. decisions about the mixed oxide (MOX) fuel program. The United Kingdom, France, and Russia currently separate the largest amounts of plutonium, and both the United Kingdom and France temporarily store plutonium for other countries, but have succeeded in reusing a relatively small amount of fuel in conventional or advanced reactors. As President Obama stated at Hankuk University in Seoul in 2012, "We simply can't go on accumulating huge amounts of the very material, like separated plutonium, that we're trying to keep away from terrorists."[7] Plutonium separation needs to be more closely aligned with use, and future fast reactors need to minimize plutonium production.

Ironically, existing technology holders are more likely to argue for retaining national control of their technology than are those yearning to acquire the technology. There are several reasons. First, sensitive nuclear activities like uranium enrichment and spent fuel reprocessing were originally developed within nuclear-weapons programs and then transferred to the civilian sector.[8] The role of government in this process has always been critical, and decisions about enrichment and reprocessing have been subject not just to market drivers but also to political factors. The example of the United States Enrichment Corporation (USEC) illustrates how political decisions can trump economic reality. Even in non-nuclear-weapon states, these elements of the fuel cycle operate under government restrictions if not outright government ownership.

Current technology holders have multiple reasons for wanting to continue national control—including security of supply, return on sunk costs, support for existing military programs (if not for production for weapons), protection of sensitive nuclear technology, and avoidance of disruption of the existing uranium enrichment services market. Prestige is likely to be a factor for some countries, such as Japan—the only non-nuclear-weapon state to operate both uranium enrichment and spent fuel reprocessing plants—which is still recovering from the Fukushima disaster.

On the front end, the argument for security of supply is undercut by the global nature of the enrichment services market—even countries that do their own enrichment also use foreign enrichment services. On the back end, no country has argued that it needs reprocessing to secure fuel for decades, nor will that happen unless uranium truly becomes scarce and a country is dependent on plutonium fuel for its nuclear power reactors.

6. Ux Consulting Company, "World Enrichment Nameplate Capacity," May 13, 2013.
7. Barack Obama, "Remarks by President Obama at Hankuk University," March 2012.
8. Exceptions include Japan, the Netherlands, and Germany.

Recuperating sunk costs is another powerful rationale for continuing to operate enrichment or reprocessing plants, but not necessarily for retaining national control. Such plants might in fact find direct foreign investment attractive. Keeping enrichment or reprocessing plants national makes sense in only two cases: if capacity is limited to that country's fuel or spent fuel management needs, and if the plants are uncompetitive with other service providers. In the latter case, a government might find that multinational ownership or operation improves competitiveness.

Retaining national control over enrichment and reprocessing facilities to support existing military programs (and potentially, in the case of Pakistan and India, to support nuclear-weapons production programs) inflames the perception that discrimination is inherent in the NPT. Keeping military production options open, whether for weapons, naval fuel, or space uses (e.g., plutonium-fueled radiothermal generators), walks back some of the progress made by the five nuclear-weapon states on voluntary offer safeguards and blurs the distinction between military and civilian uses of nuclear energy, with negative implications for the nonproliferation regime. The U.S. government's decision to transfer the Department of Energy's inventories of depleted uranium to the financially troubled USEC, in part on national security grounds, is unhelpful in this context.

On the other hand, national ownership of enrichment or reprocessing facilities may make it easier to protect sensitive nuclear technology, although corporate entities may have their own reasons to protect proprietary information. Finally, the desire not to disturb the existing uranium enrichment services market may be appealing to current technology holders, but this is countered by examples where governments have intervened in various ways to support their indigenous efforts and otherwise prop up uneconomic activity.

From the perspective of those countries that would like to acquire a national enrichment or reprocessing capability, only one of the reasons given by current technology holders is applicable: security of supply. As already noted, security of supply depends not just on national E/R facilities, but on a host of other capabilities and resources. Moreover, security of supply is better (and more cheaply) achieved through diversity of service providers.

Countries interested in acquiring a national E/R capability may also be motivated by prestige and sovereignty and possibly by the desire to develop a latent nuclear-weapons capability. The benefits of prestige and sovereignty would have to be weighed against the high costs, waste, and technical difficulty associated with national acquisition of E/R capabilities. The desire to develop a latent nuclear-weapons capability may certainly be a motivating force, but that is precisely the argument against national capabilities.

The nuclear industry is pulled in two directions—toward globalization of the supply chain but also toward consolidation of the major suppliers because of a long hiatus in growth. It is difficult to see how any emerging new suppliers will be able to compete with established players like Urenco or with vertically integrated, state-supported entities like Areva and Rosatom. New suppliers in China, South Korea, and India are all likely to have significant state support for their export activities. This increasingly national and integrated supply works against the interests of new nuclear states, making them more, not less, dependent on a single supplier.

Alternatives to national control of E/R facilities as discussed here are aimed at better securing the technology, not limiting access to nuclear energy. New arrangements for enhanced participation among states and commercial entities would not deny access to the peaceful uses of nuclear energy but would mitigate proliferation risks, strengthen certain areas of the market, and help create a uniform standard among key suppliers and vendors.

What Can Be Influenced?

Given that enrichment and reprocessing are here to stay, the modalities of their operation might offer leverage in mitigating proliferation and security risks.

Operational control of a facility: Where a state has direct control of the operation of a facility, it is in a position to change the facility's output, including making any necessary modifications to the facility itself, very quickly. A project operated by a multilateral organization and located on international territory would have the highest barriers to misuse. Operational control differs from ownership (defined as assuming the financial and political risk of building and operating the facility, selling its product or services, and benefiting from its successful operation.)

In principle, the more removed the operators are from the state, the greater the practical challenges to misuse while operators gain familiarity with the existing technology and modify equipment. The seizure of a commercial facility would be immediately obvious, allowing an opportunity for intervention.

Access to technology. If the technology is in the possession of the state or a company it controls, the state may more easily misuse the facility. Supplied on a "black-box" basis, misuse of that technology is harder and may take longer. Host state nationals will likely still have some access to the technology and protection of commercial proprietary information is not identical to protection of proliferation-sensitive information.

Legal arrangements could inhibit a state from taking control of a facility. For example, if the facility is operating under treaty arrangements, international pressure could deter a state from violating its obligations.

Technical features of the facility or the related fuel cycle could help reduce the risk of misuse, although not prevent it. For example, some products (plutonium mixed with fission products) might pose greater challenges to terrorists, although probably not to states.

The *physical location* of a facility (primarily, the state in which it is situated) is relevant for assessing proliferation risk, but it is also speculative and variable. Fuel cycle capabilities in some regions may be riskier than in others. For many countries, location is an unacceptable evaluative criterion for fuel cycle decisions because of its inherent subjectivity.

2. CURRENT NUCLEAR FUEL CYCLE MARKET STRUCTURE AND DYNAMICS

FROM AN INDUSTRY PERSPECTIVE, EACH SEGMENT OF THE MARKET IS DISTINCT FROM THE OTHERS (SEE APPENDIX), BUT VIRTUALLY ALL HAVE ONE CHARACTERISTIC IN COMMON: THE GLOBAL REACH OF SUPPLIERS. The exception is on the back end (storage and disposal).

The fuel cycle services industry is dominated by a handful of global providers who tend to have long-standing relationships with utility clients in several countries. Most sectors of the nuclear fuel cycle market function as oligopolies, characterized by concentration of 80 percent or 90 percent of the market in the hands of four or five suppliers. Barriers to entry are high because of necessary economies of scale, limited access to expensive and complex technology (particularly in enrichment and reprocessing), and the ability of existing suppliers to discourage competition (for example, by offering loss leaders to persuade a country to commit to a particular reactor technology). The strong role of governments in choosing nuclear power reactors reinforces the tendency toward oligopolistic market behavior. For example, the announcement by the United Arab Emirates government that it would eventually purchase 10 reactors all of the same type put enormous pressure on reactor vendors to engage in strategic competition to win the first bid.

Competition tends to be fierce in part because the market in most countries is static, with limited growth prospects, and because the costs of staying in the business are high. For nuclear reactor vendors, profitable fuel cycle services—particularly enrichment and fabrication—underwrite other important activities, such as reactor sales. This is crucial for private commercial companies such as Westinghouse and General Electric but also for state-backed enterprises like Tenex (Russia) and Areva (France), which are under increasing pressure to cut losses and shore up their capital structures.[1]

1. Such strains became highly visible in December 2011, when Areva announced it was halting or delaying work on a number of fuel cycle projects after operating at a loss for five years. (The company had recently undergone a change in its senior management including a new CEO.) Only a month earlier, Rosatom had announced a major reorganization, in part to offset the costs of supporting 10 "closed" atomic cities, but also to strengthen its marketing presence overseas. Apart from conversion and enrichment marketing, which will likely remain under Tenex because of the latter's well-established brand name, Rosatom's foreign

Utilities consider several factors when choosing a fuel services provider, including price, supplier reliability, supplier country diversity, process and currency diversity, proximity to enrichers, political stability, and the regulatory and competitive environments.[2] Many have relationships with providers that go back years if not decades. Custom and geography also influence decisions. For example, a U.S. utility might feel uncomfortable contracting for uranium from Areva because of concern about supply disruptions from Areva mines in Kazakhstan or Niger, whereas a European utility might be less concerned. Tenex tends to dominate markets that were supplied by the former Soviet Union during the Cold War. However, Westinghouse has secured a small portion of the fabrication market in Ukraine, on Russia's doorstep, and Tenex is grabbing a higher portion of the U.S. enrichment market, which would have been inconceivable during the Cold War.

From a recipient's perspective, diversity in suppliers is essential. Yet the combination of barriers to market entry and strong government involvement to ensure control of sensitive technology has driven a trend away from diversity of supply. The potential advantage to recipients of strong linkages between reactors and fuel fabrication and of long-term contracts for enrichment is strengthened reliability of suppliers.

While the market is global and companies have increasingly sought partners and joint ventures for investment, controls on technology have largely restricted the development of multinational enrichment and reprocessing facilities, with the exception of those in Europe. Urenco and Eurodif demonstrate that multinational commercial efforts can be successful and profitable, but there are no analogous commercial success stories for multinational reprocessing. This may be the result of poor market economics in the mid-1970s or changing attitudes about the necessity for reprocessing.

At the two ends of the fuel cycle—uranium mining and geological repositories—there are significant incentives for multinational arrangements, although for different reasons. There is a plethora of joint ventures and overseas investment in uranium mining, but political constraints have limited development of multinational arrangements for repositories.

The current nuclear fuel cycle structure has satisfied most demands for fuel cycle services apart from waste storage and disposal. And yet, new nuclear states appear to be interested in more comprehensive fuel cycle services that encompass the back end, which suggests an opening for new arrangements.

---

operations will be merged under an umbrella organization, Rosatom Overseas, with the goal of gaining a 25 percent market share of all new builds worldwide by 2030. As Rosatom Deputy CEO Kirill Komarov told a conference in Moscow in September 2011, "By 2030 we see the opportunity to participate in one degree or another in building and [maintaining] more than 70 reactors around the world." Winning new reactor projects is a key to Rosatom's survival—not because these projects will produce major income streams (indeed, they are likely to be money-losers, particularly the early ones), but because they will lead to future earnings from the provision of fuel cycle services. Reactor projects might arguably be considered a cost of being a front-end fuel services provider.

2. Ganpat Mani, "Navigating to the Land of Secure Conversion Supply" (presentation at the World Nuclear Fuel Market Conference, Seville, Spain, June 2011).

Despite spending more than $14 billion since 1982, the United States has not fully constructed or opened the spent fuel repository at Yucca Mountain, Nevada, as mandated by Congress, and the future of the project is uncertain.

Finally, the nuclear industry and market are heavily regulated because of the dual-use nature of nuclear technology. Yet technologies that are key to the development and manufacture of nuclear weapons—in particular enrichment and reprocessing—have been publicly available for decades, although this fact is not widely appreciated. One U.S. nonproliferation expert has argued that it is possible to design and build a centrifuge plant to produce highly enriched uranium without importing the technology or major pieces of equipment.[3] Laser enrichment, which has yet to be proven commercially viable, poses an additional risk because the technology requires only a small building that is indistinguishable from other industrial facilities. In addition, some doubt the efficacy of black-boxing laser enrichment technology. Thus, an approach to protecting technology that focuses on export controls may not be as useful going forward as it may have been in the past, and may need to be supplemented.

Among suppliers, two distinct styles of operation have emerged: a single giant state-backed corporation that controls all fuel cycle activities (with the possible exception of spent fuel disposal), and a loosely knit group of private, independent companies providing one or more fuel cycle services.

3. R. Scott Kemp, "Centrifuges: A New Era for Nuclear Proliferation," Nonproliferation Policy Education Center, June 5, 2012, www.npolicy.org/article.php?aid=1183&rt=&key=scott.

The use of lasers to separate isotopes of uranium (for enrichment) is a technique that has been long known, but never fully commercializable for several reasons. The current project by General Electric and Hitachi in the United States, which uses molecular laser isotope separation, could change this, with tremendous economic and proliferation implications.

State-backed fuel cycle service providers generally enjoy more political and financial support than their counterparts in the private sector. For example, Areva revealed that despite its losses it had been able to access 67 percent of a 10 billion capital spending program from external sources, which is high by private industry standards.4 (Of course, that also has left the company saddled with debt.)

At the same time, state-backed Rosatom and Areva behave in many respects like their shareholder-dominated competitors in the way they run their businesses and aggressively compete for contracts. And they are not averse to public-private partnerships. For example, Areva has teamed up with Japan's Mitsubishi to develop and sell reactors and with Shaw in the United States to build a MOX fuel fabrication plant. However, they remain backed by their respective governments either through equity stakes or direct government control. In the private sector, companies like Westinghouse and General Electric have teamed up with behemoths Toshiba and Hitachi in Japan to increase their financial and marketing muscle not only in the reactor market but in what for them is the most lucrative part of their business: providing services such as fuel fabrication and, possibly down the road for GE Hitachi, laser enrichment. These companies also maintain close relations with their respective governments and provide nuclear services for them on a contract basis.

Somewhere between these two extremes is the model followed by the Japanese and South Koreans. In these countries, nuclear providers have grouped together to boost the chances of winning reactor export business. Such contracts may contain provisions for fuel cycle services, reinforcing the trend toward an integrated approach in the fuel cycle sector.

---

4. Phil Chaffee, "Corporate: Areva Rethink Long in Coming," *Nuclear Intelligence Weekly*, December 19, 2011.

These companies work very closely with their governments, and their export credit agencies, to promote nuclear exports.

The nuclear fuel cycle market may look considerably different a few decades from now as the industry continues its consolidation with multiple partnerships between private and government-backed organizations in all areas of the fuel cycle. The big shift in energy supply and demand from Europe to Asia will also reverberate in nuclear energy, with a marked increase in nuclear programs in countries outside the traditional U.S./Western sphere of influence. This also means increased government involvement in fuel cycle activities and in nuclear programs in general. These trends suggest that there may be an opportunity to shape fuel cycle services in a way that better serves nuclear security and nonproliferation goals.

At the front end of the fuel cycle, suppliers and recipients alike take a mix of approaches to ensuring the security of supply. Nuclear generation companies and governments invest in uranium mining and milling in other states or form joint partnerships; companies may support less profitable activities like conversion through other fuel cycle activities; and many recipients contract with multiple service (e.g., enrichment) providers to ensure diversity of supply. No state except Russia has a purely national front end of the fuel cycle,[5] although enrichment suppliers have a higher share of government support, control, and ownership than suppliers in other sectors.

The lukewarm attitude of most recipient states toward supply assurances on the front end is partly based on their mistrust of solutions that leave the existing supply structure in place, even though the fuel market historically has functioned

5. Russia relies on some imported uranium because of the scale of its nuclear enterprise, but could autonomously operate the fuel cycle at a smaller scale if required.

In an attempt to convert surplus military plutonium into commercial reactor fuel, the Mixed Oxide (MOX) Fuel Fabrication Facility has been under construction at the Savannah River Site in South Carolina since 2007. Large cost overruns and an uncertain market for the specialized fuel have left the future of the project in doubt.

efficiently. The question is whether institutional approaches can (a) offer enough incentives for states to accept fuel cycle restrictions or (b) offer enough controls so that no fuel cycle restrictions are needed.

With respect to reprocessing, the significant costs of building facilities provide an incentive to states to seek outside investment. While there are probably some legal hurdles to widening international investment in what have traditionally been government-owned operations, there are precedents (e.g., Eurochemic and United Reprocessors Group [URG]) for alternatives and there are economic and political incentives on both supplier and recipient sides for widening ownership.

There is no functioning international market for disposal services at present, even though demand exists and the economics of a regional or multinational repository are encouraging. Some countries lack the appropriate geology to construct a repository, and for virtually all countries the politics of hazardous waste make siting a repository an extremely challenging undertaking. As a result, users would likely pay a healthy premium for the provision of such a service. Despite the obvious economic advantages and the potential safety and security benefits, several obstacles have prevented realization of concrete multinational back-end projects as yet. Some of the key challenges include siting, transportation security and costs, waste acceptance criteria, and security and nonproliferation.

The nonexistent back-end market potentially offers opportunities for leveraging the provision of a spent fuel disposal pathway in return for a commitment not to pursue sensitive fuel capabilities such as enrichment and reprocessing. According to a 2007 U.S.-Russian National Academies workshop, "arrangements that would provide assured return of spent nuclear fuel could provide a much more powerful incentive for countries to rely on international nuclear fuel supply than would assured supply of fresh fuel, because assured take-back could mean that countries would not need to incur the cost and uncertainty of trying to establish their own repositories for spent nuclear fuel or nuclear waste."[6] Completed contracts with Russia to supply fresh fuel and take back used fuel as part of reactor purchases are indicators of interest in these arrangements for new nuclear energy states.

---

6. U.S. Committee on the Internationalization of the Civilian Nuclear Fuel Cycle, Committee on International Security and Arms Control, National Academy of Sciences, and National Research Council, *Internationalization of the Nuclear Fuel Cycle: Goals, Strategies, and Challenges* (Washington, DC: National Academies Press, 2009), 48, http://www.nap.edu/openbook.php?record_id=12477.

3. PAST PROPOSALS TO REDUCE RISKS

ALTERNATIVES TO "NATIONAL" FUEL CYCLE CAPABILITIES CAN RANGE FROM MINIMALIST OR INCREMENTAL APPROACHES TO A COMPLETE REVAMPING OF THE SYSTEM. Most proposals have tended to be minimalist, focusing on adding assurances to the existing structure. A few have been more far-reaching.

In reducing proliferation risks, conventional wisdom is that a combination of technical and institutional approaches is needed. While the science and technology community conducts research and development to enhance "proliferation-resistance," particularly of a closed fuel cycle, the proliferation policy community often states that proliferation-resistance is of limited value and suggests strongly that institutional measures provide the only prospect for meaningful progress.

At the front end of the fuel cycle, current practice among technology holders is to provide enrichment equipment, if at all, on a solely "black-box basis." This technical approach varies among suppliers: Urenco has gone so far as to cordon off the technology into a separate company, the Enrichment Technology Corporation (ETC), and although Areva has a share in ETC, it does not have access to Urenco-origin centrifuge technology. There are several practices that Urenco engages in to protect the technology, including creating free-trade zones for importing centrifuges to another state (like the United States). Less is known about Russia's practices to "black box" the technology, but since Russia has only sold enrichment equipment to China, there is little experience so far. No enrichment supplier has set up new plants in states other than nuclear-weapon states (Russian technology in China and ETC technology in the United States and France).

At the back end of the fuel cycle, Areva has plans to sell only co-extraction of actinides (COEX) technology, which does not result in a separated plutonium product.[7] Expert discussions have been fairly pessimistic about the prospects

7. COEX separates a homogenous uranium-plutonium mixture. While this places the demand of an additional step to access pure plutonium for would-be proliferators, the technical barrier is not significant.

The Role of the European Supply Agency

The Euratom Treaty, signed in 1957 by six countries, created the European Atomic Energy Community. Members now include all members of the European Union (EU). The treaty created two agencies, the European Supply Agency (ESA) and the Euratom Safeguards Office. Under the treaty, "the supply of ores, source materials and special fissile materials shall be ensured . . . by means of a common supply policy on the principle of equal access to sources of supply." The ESA has a

> right of option on ores, source materials, and special fissile materials produced in the territories of Member States and an exclusive right to conclude contracts relating to the supply of ores, source materials and special fissile materials coming from inside the Community or from outside. The Agency may not discriminate in any way between users on grounds of the use which they intend to make of the supplies requested unless such use is unlawful or is found to be contrary to the conditions imposed by suppliers outside the Community on the consignment in question.

The ESA manages its responsibility to ensure sufficient supply of nuclear materials in a few different ways; it promotes the view that reactor operators should enter into long-term contracts with a diverse group of suppliers and that utilities should consider maintaining strategic inventories of fuel cycle materials.

The ESA monitors the fuel cycle market and approves individual purchases of fuel cycle materials. The Euratom Treaty requires the ESA to be a signatory to supply contracts if the purchaser is an EU utility or research reactor operator. The ESA is also a signatory to contracts involving the sale of nuclear material within the European Union, or in cases where there are exports from, or imports to, companies in the EU. This requirement applies to goods, not services (as defined by the European Court of International Justice). This means that the ESA is party to fuel cycle contracts involving natural uranium or enriched uranium product, but not to contracts for services such as uranium conversion or uranium enrichment.

for technical approaches to reduce proliferation risks on the back end, particularly since separations technology is well known and not very difficult technically.

There have been many proposals for institutional approaches to mitigate fuel cycle risks. The intent here is not to catalog them, but to understand their objectives and distinguishing features.[8] The discussion here is divided into three parts, each associated with a different sector of the sensitive fuel cycle. Broadly speaking, emphasis has shifted over time

---

8. For an overview of existing proposals, see Yury Yudin, *Multilateralization of the Nuclear Fuel Cycle: Assessing the Existing Proposals* (Geneva: UN Institute for Disarmament Research, 2009), http://www.edam.org.tr/Media/Files/158/fuelbankproposals.pdf.

according to market influences, policy changes, and proliferation activity. Thus, we saw more emphasis on the threat of expanded reprocessing in the 1970s but on expanded enrichment in the 2000s. In any case, there is no technical "silver bullet" for the challenges and any effective solution must include a combination of institutional and legal frameworks in addition to technical approaches.

Front End of the Fuel Cycle

Most proposals for alternatives to current fuel cycle arrangements embellish the current system by adding layers or backups to existing capabilities. A few create new structures or concepts. Almost all of the proposals based on creating fallback options for consumers in the commercial market have been championed by supplier states and perceived as chiefly concerned with minimizing market disruptions while maintaining market share. In contrast, virtually all the proposals that advocate significantly restructuring the market come from individuals or organizations without government affiliation. Among the proposals:

- The *World Nuclear Association's Three Tier Concept* calls for existing enrichers, with government and IAEA support, to collectively guarantee supply in the event of a politically motivated fuel cutoff.[9]
- Under the *UK Nuclear Fuel Assurance* ("enrichment bonds") proposal, supplier governments would agree to surrender the right to withhold export approval in favor of an IAEA final decision.
- The *International Low-Enriched Uranium (LEU) Fuel Bank* was approved by the IAEA Board of Governors in 2010 and is intended to be a stockpile of last resort that the IAEA can release to provide fuel for a civil nuclear power reactor whose fuel supply has been disrupted and for which a commercial replacement is unavailable. It is intended to provide a backup to the fuel market, without disrupting that market, so that states do not need domestic enrichment capacity for nuclear energy security.
- The *U.S. International Framework for Nuclear Energy Cooperation* seeks to establish international supply arrangements for reliable, cost-effective fuel services and identify other areas of global infrastructure that could be improved through international cooperation within the current market framework.
- Russia's *Global Nuclear Power Infrastructure* envisions three to five international enrichment centers using black-box Russian enrichment technology. The International Uranium Enrichment Center in Angarsk is the first and only step thus far toward realizing this vision.

---

9. The Six Country Concept (France, Germany, Netherlands, Russia, United Kingdom, and United States) complements the Three Tier Concept by removing the requirement that suppliers contribute equal shares and introduces the option of suppliers transferring third-tier backup reserves to the International Atomic Energy Agency.

- The *Terms for Reliable Uranium Service Transactions through Leasing* envisions a trust consisting of private and public investment banks, private nonproliferation entities, and high-net-worth individuals negotiating supply contracts with suppliers. Essentially a mechanism to enable the existing market to function more efficiently, it is based on the expectation that pooling nuclear fuel purchases and attendant services would produce a competitive market for emerging nuclear energy states and other lessees while suppliers could more easily manage their supply decisions with a predictable demand.[10]

- The Nuclear Islands concept advocates creating an *International Nuclear Fuel Cycle Association* that would initially encompass all uranium enrichment activities within "internationally secured leased areas" and later be extended to cover other sensitive fuel cycle activities. This body is intended to augment and support the IAEA. Its members (suppliers and customers) would be required to do business only with other members.[11]

- An *International Nuclear Fuel Agency* with authority over all national enrichment facilities was proposed by the Stockholm International Peace Research Institute in 1979. Laser enrichment and plasma separation would be stopped, centrifuges would be phased out, and gaseous diffusion and chemical-exchange technology would be employed in an effort to use only technology that presents serious obstacles to national construction beyond acquiring classified data. Membership would not include a withdrawal provision, and the agency would be empowered to enact sanctions against violators.[12]

- The *Multilateral Enrichment Sanctuary Project,* proposed by Germany, suggests that a new, multilateral enrichment facility (or facilities) be built and operated by an international company (or companies) and administered by the IAEA on territory ceded to the agency by a host nation.

- Under the *IAEA Standby Arrangements,* proposed by Japan, all states with the ability to supply front-end services would annually furnish the IAEA with their supply capacities, from uranium supply to fuel fabrication. Armed with this information, the IAEA would act as an intermediary between consumers and suppliers should a supply disruption occur. The Euratom Supply Agency, which acts as a supranational authority to ensure security of supply, has practiced a hybrid approach for many years. Deriving its authority from the Euratom Treaty and the establishment of the European Economic Community (i.e., the Common Market), it aims "to ensure a regular and equitable supply of nuclear fuels to EU users. To perform this task, the agency applies a supply policy based on the principle of equal access to sources of supply."[13] Thus, at least in Europe, there is a precedent for supply organizations with significant authority. There are no restrictions on additional

10. Stephen Goldberg, James Glasgow, and James Malone, "'TRUST,' an Innovative Nuclear Fuel Leasing Arrangement" (presentation to the World Nuclear Fuel Cycle Conference, Budapest, Hungary, April 2007). The authors cite the 1994 HEU Purchase Agreement between the United States and Russia, accomplished in a budget-neutral fashion, as an excellent model.

11. Christopher E. Paine and Thomas B. Cochran, "Nuclear Islands: International Leasing of Nuclear Fuel Cycle Sites to Provide Enduring Assurance of Peaceful Use," *Nonproliferation Review* 17, no. 3 (2010): 452, http://cns.miis.edu/npr/pdfs/npr_17-3_paine_cochran.pdf.

12. Allan Krass et al., *Uranium Enrichment and Nuclear Weapon Proliferation* (London: Taylor & Francis, 1983), 88–91.

13. Euratom Supply Agency, "Welcome to the Euratom Supply Agency," accessed October 16, 2014, http://ec.europa.eu/euratom/index.html.

states acquiring enrichment capabilities, but there have been no demands to do so among Euratom states.

Most of these approaches apply a market solution to a political problem. The fuel supply market functions efficiently. States that might be concerned about politically driven fuel cutoffs are not reassured by approaches that preserve the current network of supplier states and companies. Indeed, it is not clear that even the proposals advocating a restructure of the market would necessarily have an impact on the current suppliers' market control. A 2010 study sponsored by the American Academy of Arts and Sciences observed: "The emphasis on ensuring security of supply of other services, such as reactor construction, fresh fuel, enrichment, and reprocessing, is misplaced. All of these services are supplied commercially at present, and a customer country currently has a choice of suppliers that may well be wider than would result from implementation of initiatives that create a two-tier system of nuclear supplier and user countries."[14] Unless there is a broader consensus on the problem to be solved, market-based solutions are not likely to find traction despite their potential value.

Back End of the Fuel Cycle: Reprocessing and Spent Fuel Disposal

There have been fewer proposals aimed at the backend of the fuel cycle. First, reprocessing technology is fairly simple and well known. Second, the recent exposures of the A.Q. Khan network's exports to Iran, North Korea, and other states and Iran's clandestine enrichment plants have focused public attention on enrichment (even though Iran has also engaged in some undeclared back-end activities). More fundamentally, however, in an era of relatively plentiful uranium at reasonable prices and a fuel cycle that relies predominantly on LEU for light water reactors, a cutoff in enrichment could have an impact on fuel supplies, while a cutoff in reprocessing services would have little practical effect on fuel supplies, unless plutonium becomes widely used in fast reactors.

Reprocessing has few technical secrets, but virtually all reprocessing facilities are owned in whole or part by national governments, except in Japan. In the 1970s, international collaboration in both enrichment and reprocessing was attractive because of the small domestic markets, long lead times to develop technology, technical complexity, and cost. For example, Eurochemic, which operated from 1966 to 1974, involved significant technology development and sharing among the members. In contrast, United Reprocessors sought to

14. Charles McCombie et al., *Multinational Approaches to the Nuclear Fuel Cycle* (Cambridge, MA: American Academy of Arts and Sciences, 2010), 8, http://www.amacad.org/multimedia/pdfs/publications/researchpapersmonographs/isaacsInside.pdf.

maximize collaboration among national plants in the United Kingdom, Germany, and France. Although some technical information was shared, United Reprocessors operated more like Urenco than Eurochemic, requiring agreement among all three states to share information with non–United Reprocessors states.

With the decades-long downturn in nuclear energy, however, there have been few proposals to multinationalize recycling. International consortium/IAEA-related arrangements to provide assurances of reprocessing capacity do not currently exist for two main reasons. First, fewer states are sending their spent nuclear fuel (SNF) across national borders for reprocessing because of policy decisions or high costs, and because in most cases the resulting wastes will ultimately be shipped back, limiting the long-term added value of reprocessing. Second, states starting out with nuclear energy typically approach the issue of waste with a wait-and-see attitude because they can store fuel for many years.

Since reprocessing efforts are all government-owned or conducted by companies controlled by governments, intermediate approaches such as purchasing shares in such efforts are likely to be more difficult in the back end than the front end. Practically speaking, reprocessing service providers are much more likely to simply offer their services than to offer an ownership share. The 2005 IAEA report on Multilateral Approaches to the Nuclear Fuel Cycle stated that there "will be sufficient reprocessing capacity globally for all expected demands for at least two decades. Therefore, objectives of assurances of MOX supply can be fulfilled . . . without [multilateral nuclear approaches] involving ownerships."[15] This statement is likely predicated, however, on the assumption that ownership would entail building new facilities, rather than purchasing shares in existing facilities.

Excess capacity at existing plants that reprocess foreign spent fuel makes building a new facility costlier than seeking already-existing services. Compared to plans to develop a new national facility, however, there could be economic benefits to distributing the investment burden. Here, the example of Eurochemic may be instructive.[16] An initial impetus for proceeding with multinational reprocessing among the Organization for Economic Cooperation and Development (OECD) states was the fact that a reprocessing plant would require higher capital costs than other nuclear projects (such as an enrichment plant or heavy water production plant). However, the final economic outcome for eight years of reprocessing was "frankly disappointing."[17]

15. International Atomic Energy Agency (IAEA), *Multilateral Approaches to the Nuclear Fuel Cycle: Expert Group Report to the Director General of the IAEA* (Vienna: IAEA, 2005), http://www-pub.iaea.org/MTCD/publications/PDF/mna-2005_web.pdf.

16. Jean-Marc Wolff, *Eurochemic: European Company for the Chemical Reprocessing of Irradiated Fuels, 1956–1990* (Paris: OECD, 1996), 311, http://www.eurochemic.be/nl/documents/68-eurochemic-EN.pdf.

17. There were many reasons for this, including the loss of political cohesion among the 13 partners and overcapacity in reprocessing. Competing plants (and agendas) in France and Germany sealed the fate of Eurochemic by the early 1970s. Eurochemic's most significant contributions have been described as helping facilitate the development of reprocessing from first-generation military efforts to second- and third-generation commercial plants, and more disconcertingly, providing international training in reprocessing.

FUEL LEASING/TAKE-BACK

Proposals on fuel leasing/take-back arrangements for small or emerging nuclear programs are both a powerful sales incentive and an effective way to strengthen nonproliferation and safety. Take-back is central to Russia's nuclear reactor deals with Iran, Turkey, and Vietnam. The fuel leaser does not necessarily have to dispose of the returned spent fuel; it could be sent (for example, through an IAEA-brokered deal) to a third-party state or a regional or multinational fuel cycle center located elsewhere. For nuclear newcomers, the provision of bundled or cradle-to-grave services from full-service companies or consortia might be attractive because it is simpler and removes the need for negotiating multiple contracts in an unfamiliar market. However, for more established utilities, conversion, enrichment, and fuel fabrication contracts are negotiated separately and on different timetables. Operators are constantly making small tweaks to fuel designs that increase reliability; these tend to mature in two- to three-year cycles. This means that utilities prefer to have their fuel fabricated at the very last moment before delivery to receive the most efficient fuel possible, so there is not much incentive to get locked into long-term fuel contracts.

Fuel leasing proposals have proved less appealing to the customer when the spent fuel is only being removed for reprocessing and the remaining high-level waste will eventually be returned. This is currently the case with the French and British reprocessing programs, although earlier contracts left the wastes at the reprocessing plants. While the U.S. Global Threat Reduction Initiative takes back spent fuel from research reactors worldwide to reduce proliferation dangers, the U.S. Congress has thus far shown no interest in doing the same for civilian power reactors.

MULTINATIONAL/INTERNATIONAL REPOSITORIES

A 1987 preliminary study by the NEA weighed two internationalization paths: the creation of a dedicated international repository, and opening an existing national repository to accept material on a commercial basis from other states. While the latter was determined to be the more realistic option, the lack of progress on national repositories convinced the authors that a more comprehensive study would be premature.[18] The 2001 Joint Convention on the Safety of Spent Fuel Management and on the Safety of Radioactive Waste Management counseled that "in certain circumstances, safe and efficient management of spent fuel and radioactive waste might be fostered through agreements among Contracting Parties to use facilities in one of them for the benefit of the other Parties, particularly where waste

18. Nuclear Energy Agency, *International Approaches to the Use of Radioactive Waste Disposal Facilities: A Preliminary Study* (Paris: Organization for Economic Cooperation and Development [OECD], 1987).

The Waste Isolation Pilot Plant (WIPP) in eastern New Mexico has been disposing of transuranic waste from US military programs since 1999. A February 2014 radiation leak forced the closure of the nation's first and only nuclear waste repository; it is not clear when the facility will reopen.

originates from joint projects."[19] In 2004, the IAEA published a study examining scenarios of cooperation for the development of multinational repositories.[20]

Past proposals have included concepts for a shared repository that is developed by a group of states. The model of the 1980 Low-Level Radioactive Waste Policy Act in the United States is that several states with small but developed nuclear programs would be motivated by economies of scale and agree to use a shared repository.[21] From 2003 to 2008, the European Commission funded pilot studies on the feasibility of shared regional storage facilities and geological repositories in Europe and options for the establishment of a European Repository Development Organization (ERDO).[22] In late 2003 and early 2004, experts from Austria, Bulgaria, Croatia, the Czech Republic, Hungary, Slovakia, and Slovenia met to discuss regional High Level Waste (HLW) disposal solutions, so that, as states with small amounts of radioactive waste, they could ensure that one of them acquires the necessary

---

19. IAEA, "INFCIRC-546: Joint Convention on the Safety of Spent Fuel Management and on the Safety of Radioactive Waste Management," December 24, 1997, http://www.iaea.org/publications/documents/conventions/joint-convention-safety-spent-fuel-management-and-safety-radioactive-waste.

20. IAEA, *Developing multinational radioactive waste repositories: Infrastructural framework and scenarios of cooperation* (Vienna: IAEA, October 2004), http://www-pub.iaea.org/MTCD/publications/PDF/te_1413_web.pdf.

21. The Act "encouraged states to form regional compacts to meet their collective disposal needs, minimize the number of new disposal sites, and more equitably distribute the responsibility." After three decades of interstate conflict and numerous court cases, the first disposal facility under the Act opened in Andrews County, TX, in 2011. The Act has also had the unexpected benefit of functioning as "an essential vehicle of protection for existing host states to exercise authority over [low-level radioactive waste] management." Daniel Sherman, *Not Here, Not There, Not Anywhere: Politics, Social Movements, and the Disposal of Low-Level Radioactive Waste* (Washington, DC: Resources for the Future, 2011), 189.

22. The ERDO Working Group documents can be found at www.erdo-wg.eu/Documents.html. While a national declaration of willingness to be a repository host is not necessary to join the ERDO Working Group, membership is conditional: countries cannot rule out the possibility of hosting a repository, and the consortium's work cannot be allowed to stop a national repository program. Sweden and Finland have expressed some concern that ERDO's work will undermine their programs.

technology and institutional structures for a repository.[23] The UAE and the Gulf Cooperation Council have also supported initial studies on regional cooperation at the back end.

States seeking disposal pathways beyond their borders require assurance that spent fuel and HLW will be removed in a reliable and timely manner for as long as it is produced. These arrangements require the siting of waste facilities, and accepting foreign spent fuel would likely further complicate this process. Such facilities might be more acceptable if they advanced the international community's security and nonproliferation goals and offered substantial financial gain. The successful siting, construction, and operation of the Waste Isolation Pilot Plant for transuranic waste in New Mexico, the first facility of its kind in the world, demonstrates that local communities and states can be enthusiastic partners in the right circumstances.

About half of the countries currently operating nuclear power plants have fewer than five plants, which means that they do not accumulate spent fuel very quickly. It also means that spent fuel management is proportionately a very expensive component of these countries' nuclear programs. Access to a disposal pathway that removes all responsibility for spent fuel management may make more financial and political sense. Even countries with larger programs may find economic, security, and political benefits in multinational approaches.

Storage and Disposal

Provided a longer-term multinational or regional disposal strategy is also in place, multinational interim storage offers the nonproliferation benefit of enabling spent fuel to be removed more quickly from numerous small stores to a centralized, well-safeguarded location. A study on interim storage envisioned a commercial storage facility for international customers designed to store up to 10,000 metric tons of spent nuclear fuel (both legacy and future spent fuel) in dry casks for up to 100 years. Short-term (i.e., 20-year) storage contracts would allow customers to take advantage of new spent fuel management developments.[24] The window to develop long-term solutions can also be expanded in some locations by increasing on-site storage.

Interim solutions are hampered by a lack of trust that permanent solutions will

23. Charles McCombie and Neil Chapman, "Nuclear Fuel Cycle Centres—an Old and New Idea" (presentation at the Annual Symposium of the World Nuclear Association, London, England, September 2004).

24. Stephen Goldberg, Robert Rosner, and James Malone, *The Back End of the Nuclear Fuel Cycle: An Innovative Storage Concept* (Cambridge, MA: American Academy of Arts and Sciences, 2012), 13–18. According to the Blue Ribbon Commission on America's Nuclear Future, "After an initial period of cooling in wet storage (generally at least five years), dry storage (in casks or vaults) is considered to be the safest and hence preferred option available today for extended periods of storage (i.e., multiple decades up to 100 years or possibly more)." It should be noted, however, that the Commission did not advocate using temporary storage as a reason to delay action and strongly recommended moving forward promptly to establish interim centralized storage and repository programs. Blue Ribbon Commission on America's Nuclear Future, *Report to the Secretary of Energy* (Washington, DC: Blue Ribbon Commission on America's Nuclear Future, January 2012).

be made available, or that the interim storage will, by design or default, become permanent. Some fear that multinational programs will undermine national programs that are making progress. Given this problem, questions have been raised about the advisability of spending essentially the same amount of time, effort, and political capital on a temporary fix rather than concentrating on a permanent solution.[25] Success at operating an interim storage facility may help to build the trust that is necessary for communities, states, and federal governments to support a permanent repository. But in the current environment, this will not happen quickly or easily. Interim storage and final disposal each have their own technical, economic, and political advantages and disadvantages; neither should be pursued to the exclusion of the other.

Several obstacles have prevented realization of concrete multinational back-end projects as yet. Some of the key challenges are as follows:

HOST STATE SITING

Public acceptance remains the greatest challenge, followed by the existence of appropriate geology. The well-developed repository programs in Sweden and Finland, as well as the progress that has been made in France and Canada, have required decades to build the requisite public trust, often after early missteps, effective opposition, and other delays. Some states have failed to win sufficient trust despite decades of effort. The United States, which has spent more time and money working on the problem than any country, may not be much closer to a political solution than it was 60 years ago. The most active current multinational initiative, the ERDO proposal in the European Union, has deliberately postponed the sensitive issue of siting until much more work has been completed on both geological screening and trust-building. For multinational repositories, stable governments with independent, transparent regulators will be required.

TRANSPORT SECURITY AND COST

Regional or multinational repositories, as opposed to national repositories, require more SNF shipments over greater distances and the shipments are trans-boundary. This increases cost and security risk. But if managed properly, the relatively modest increases in short-term costs and risks are outweighed by the longer-term costs and risks posed by national repositories and/or continued on-site SNF storage.

---

25. The demand for, and economics of, multinational storage have been questioned in the European context. See Charles McCombie Chapman, N. A. Chapman, and P. Richardson, *Economic Aspects of Regional Repositories: SAPIERR II Work Package 3,* European Commission Community Research, April 2008, 25–29, http://www.erdo-wg.eu/Documents_files/SAPIERR%20II%20WP-3%20web.pdf.

Dry casks are an economic and safe method of storing spent fuel for extended periods, in order to maintain sufficient space in spent fuel pools while disposal pathways are found. Dry cask storage is used widely in the US and to varying degrees in Canada, Europe, and Japan.

WASTE ACCEPTANCE CRITERIA

Criteria for accepting waste applies to both the SNF and its packaging and must be settled in advance to avoid cost increases and safety problems. The handling facilities at the repository and also the acceptance process will be complicated by the need to incorporate legacy SNF inventories in many states.

SECURITY AND NONPROLIFERATION

Some observers believe that large volumes of SNF at one or a few sites could attract subversive attacks, particularly several hundred years down the road when the fuel has cooled sufficiently to be handled. However, a multinational safeguarded facility designed with high levels of security would present a difficult, and thus uninviting, target while small quantities of SNF at multiple locations around the world may be less secure.

LIABILITY

Ownership of the spent fuel and liabilities for cost increases and system malfunctions must be transferred to the repository operator, host state, repository partners, or some combination of these. The timing and scope of this transfer have financial implications for the customer and the host, which will require changes in national laws. A further level of complexity is added if spent fuel is treated as a resource and the repository is retrievable. Increased costs will also result from standardizing the applicable laws and regulations of the participants. The complexities of managing liability and other legal arrangements are minimized to a significant extent in the European Union and Euratom due to an existing common understanding of basic concepts and an agreed regulatory framework.

COST SHARING AND SCHEDULING

A multinational repository should be able to service several independent utilities with differing fuel types and differing disposal schedules. The partners should be able to develop a costing algorithm that is accepted as fair by all users and a timetable for disposal that suits the national requirements while allowing continuous, optimized operation of the shared facility. Some of these challenges are already shared by those developing strictly national programs.

4. A NEW APPROACH TO THE NUCLEAR FUEL CYCLE

THE FEW BARRIERS TO PROLIFERATION THAT DO EXIST ARE POLITICAL AND REGIME-RELATED, BUT THESE ARE INADEQUATE IN THE FACE OF WEAK ENFORCEMENT AND A LACK OF INTERNATIONAL CONSENSUS. Moreover, political, economic, and regional security instabilities reduce confidence that the current structures are good enough to ensure that nuclear energy is used for exclusively peaceful purposes moving forward, even if challengers to the existing regime are few.

The status quo is inconsistent with a future that recognizes the value of nuclear energy and the need to impede proliferation. There is clearly a need for an integrated approach to the fuel cycle, based on principles that apply to all aspects of the development and deployment of sensitive technologies in the front and back ends of the fuel cycle.

Objectives for Production (Enrichment and Reprocessing)

The good news is that civilian and military highly enriched uranium (HEU) production worldwide is already voluntarily limited, but it should be eliminated for civilian purposes under this new approach. Any new enrichment facilities need to meet specific criteria detailed in this chapter or the state would be subject to a general ban on nuclear-related commerce.

Likewise, new reprocessing facilities would need to meet specific criteria, including multinational ownership/operation, willingness to accept spent fuel from others, and operation scaled to consumption of separated plutonium. A fundamental objective is to have no net increase in the amount of nuclear-weapons-usable material. In research and development, the long-term goal should be a phaseout of technologies that result in separated plutonium being produced and stockpiled.

Breeding plutonium in fast reactors should be minimized via technical approaches. If possible, they should be operated in burning or equilibrium modes so that there is no net plutonium production. If plutonium breeding becomes necessary in the future, every effort should be made to avoid separation of weapons-grade plutonium. Fuel or breeding assemblies containing low burn-up plutonium should not be reprocessed separately, but in a mix with other assemblies (core fuel or thermal reactor fuel) to avoid any plutonium product at or near weapons-grade. Every effort should be made to avoid plutonium separation—for example, keeping the plutonium product mixed with actinides and fission products so that it is self-protecting against diversion or theft and requires further reprocessing before weapons use is feasible.

Addressing material production and associated technologies in accordance with these guidelines requires changes to many current practices. IAEA safeguards should be implemented on all enrichment and reprocessing (E/R) facilities, including in the nuclear-weapon states, as well as any facility processing civilian highly enriched uranium, and an Additional Protocol must be in force in all countries. Diversified partnerships should be the new norm for ownership and operations of new facilities and such arrangements can be phased in for existing facilities. These arrangements could include among others, financial investment, equity stake, and/or rights to output/product.

Third-party control of sensitive technology and processes, which could include manufacture, installation, and maintenance, such as the model established by Enrichment Technology Company (ETC) and Urenco, is needed to limit the ability of an operator to copy and build components or to modify and operate the facility without the natural learning curve. These arrangements facilitate early, credible, and unambiguous detection, enabling intervention in the case of breakout. Finally, states must make political commitments and regulatory assurances that there will be no new HEU production or civil facilities to use HEU, and no plutonium separation without a realistic path for consumption.

New reprocessing facilities need to be linked to reactors that will contemporaneously consume the separated material. No reprocessing should be undertaken unless there is a concurrently licensed repository for high-level waste. Breeding in fast reactors fueled by plutonium should be minimized through technical reactor design approaches.

Finally, operators of reprocessing facilities should be willing to accept spent fuel from others, as long as the material is appropriately regulated for quality assurance/quality control and its processing is scaled to consumption to avoid stockpiling. Reactor operators should be encouraged to consume recovered plutonium consistent with the recommendations in this report, provided reprocessing output is scaled to consumption.

Enrichment and reprocessing facilities, existing or planned, should demonstrate that they meet the following criteria for achieving nonproliferation and security objectives:

- Additional Protocol IAEA safeguards on all E/R facilities (including in the nuclear-weapon states), plus any facility processing civilian HEU/plutonium such as mixed oxide (MOX) fuel fabrication plants ; and no outstanding safeguards implementation issues before the IAEA Board of Governors

- Diversified partnerships for facility ownership and operations (immediately for new facilities and phased in for existing ones, possibly proportional to export quantities), including options such as financial investment, equity stake, and rights to output or product
- Sensitive technology and processes manufactured, installed, controlled, and maintained for new facilities by a third party (such as the model established by ETC and Urenco) that is different from the operator, in order to accomplish the following objectives:
  - Limit the host's ability to copy and build components.
  - Limit the host's ability to modify and operate without the natural learning curve.
  - Limit a proliferant state's advance planning time.
  - Ensure that detection of potential breakout is early, credible, and unambiguous, facilitating intervention.
- Political commitment and regulatory assurances of no HEU production, no new facilities licensed for HEU production or use, and no civil HEU-fueled facilities
- Political commitment and regulatory assurances of no plutonium separation unless there is a concurrently available path for consumption

Reprocessing plants to separate spent nuclear fuel into plutonium, uranium, and fission products are generally large industrial plants that emit detectable effluents, but smaller reprocessing efforts can be easier to hide. Plants are generally government-owned, although the Rokkasho Reprocessing Plant, shown above, is owned by Japanese utilities.

Objectives for Use of Fissile Material

Today, HEU can be replaced by less sensitive low-enriched uranium (LEU) in nearly all current uses, thanks to technical developments. This is true for civilian as well as naval uses. Efforts now focusing on minimizing civilian HEU should set their sights on eliminating HEU. Some of the existing stockpiles can be earmarked for use until the full technical transition is complete and the rest should be downblended to LEU. Existing programs to convert research reactors using HEU to LEU and to remove HEU from as many sites and countries as possible must be supported and expanded.

The tons of separated plutonium that exist around the world need to be disposed of or consumed. One obvious option is to mix separated plutonium with waste and bury it. Another is to leave the option open of burning plutonium in advanced reactors. In all cases, the objective here would be to have no net increase in nuclear-weapons-usable material and to use existing material before production of new material where there are conversion and consumption paths available.

Objectives for Disposal

No repositories for commercial spent fuel are currently operating anywhere in the world, although Sweden and Finland have made good progress toward establishing them. Failing to address this issue is ultimately unsustainable. The approach proposed here would use back-end solutions to influence front-end decisions—that is, it would offer access to repositories only to countries that do not pursue national fuel cycle facilities in contravention of the criteria listed below (see the Elements of a Best Practices Approach). However, this relies on the ability of states to establish such repositories. The challenges of siting repositories are legion, but may be alleviated if there is a market, and clear financial and other societal benefits may be realized by linking the front and back ends of the fuel cycle.

Several spent fuel management pathways are conceivable, none of which are mutually exclusive. Fuel leasing or take-back would enable a profit for suppliers. While an expense would be incurred if no fee is charged, providing the service would give any supplier a significant market advantage. Several variants of this service are possible. For example, the service provider may be responsible for the spent fuel. (For some full-service providers, this may also include responsibility for eventual decommissioning of the plant; for others, plant decommissioning will remain the responsibility of the user. Over time, the market advantages accruing to suppliers offering the former service might put pressure on other suppliers to do the same.) This approach still requires the supplier to develop its own spent fuel disposal pathway.

Fuel cycle parks could be expanded to provide multiple fuel cycle services, including final disposal. Collocating sensitive fuel cycle facilities and reducing transportation requirements would increase financial attractiveness for the host and provide a significant market advantage. The creation of regional or international repositories would markedly reduce any incentive to reprocess spent fuel and could be linked to a future "no new enrichment" provision. The provision of this service should be financially attractive for the host and, even with high disposal prices, should be financially attractive for users. While liability issues will need to be worked out, there is evidence that the European Union is allowing such regional approaches to evolve.

Finally, interim storage incorporating any of the above-mentioned scenarios would buy time. More countries could conceivably provide interim storage than permanent geological disposal; the fuel would be easily retrievable; a financial model exists; and there is evidence that some countries, perhaps Japan, South Korea, and Taiwan, would use this service. However, it is unclear that the public would be more accepting of this approach than of repositories, and it is not a permanent solution.

The specific steps for spent fuel include diversified partnerships for repositories, such as access and space availability for partners and consideration of diversified partnerships and regional approaches to new interim storage and/or disposal programs for spent fuel and high-level waste. In both cases, such steps should be careful not to undermine existing national programs.

Elements of a Best Practices Approach

To be economically and politically sustainable, a few steps must be supported and consistently applied by both government and industry:

- Accept and implement the most recent nuclear security guidelines from the IAEA (INFCIRC-225).
- Establish an independent national regulator according to metrics developed by third parties.
- Require, through state-to-state arrangements, prior consent for downstream use of material and by-products for new supply contracts.
- Contribute to a consumption tax to pay for additional safeguards and any other burdens on the IAEA (scaled by inspection effort required).

To be successful, a new, integrated approach to the fuel cycle must change the way existing technology holders operate and not just limit the options for new entrants. Neither can it require a new commercial or government superstructure that would wholly replace existing market arrangements.

The basic premise of the approach recommended in this report is that the risk of misuse can be reduced by removing fuel cycle facilities from national control. But this is easier said than done. Moreover, simply changing corporate structures is not enough to prevent proliferation. Instead, suppliers and recipients need to agree on a set of norms that will apply both to new activities and to existing pursuits, the latter after a transition as fuel cycle facilities follow their life cycles. The resulting system of best practices should strive for political, economic, institutional, and technical sustainability through mutually reinforcing commitments to objectives consistent with nonproliferation, security, and commercial success, as listed below.

- *Equal Access and Shared Benefits:* States should expand opportunities for other countries to invest in commercial ventures and share in the benefits, without increasing security risks.
- *Shared Costs and Burdens:* Government and industry could contribute to a consumption/sustainability tax to pay for additional safeguards and any other burdens on the IAEA. All facilities would follow established and evolving best practices for security, safety, and safeguards for all materials and facilities in use or in transit.
- *Early Detection:* The system should maximize opportunities for early detection of noncompliance, including through improving transparency.
- *Minimizing Weapons-usable Material:* The system should yield the least possible amount of weapons-usable material.
- *Market-Driven Expansion:* No new facilities capable of producing weapons-usable material should be constructed unless there is unmet commercial demand.
- *Incentives and Consequences:* Governments and industry should benefit more from adhering to their commitments than from breaking them.

Such an approach would challenge the current de facto monopoly of the fuel cycle owner-states and increase access to the peaceful benefits of nuclear energy through a layered system that includes active participation by states in facilities and fuel assurances. Current technology holders would have to accept new partners and regulatory structures, so the changes to the system would not be disproportionately felt by nontechnology holders. In order to make the decision to support a change easier for any entity, state or commercial, new approaches to the operation of fuel cycle facilities should also address concerns about both security of supply and equity.

Applying Best Practices

These best practices should not be controversial, but applying them to fuel cycle activities will require changes to current arrangements. An approach that would meet these criteria would address production, use, and disposal of fissile material and waste. Implementation in all these areas will form the core elements for a new approach.

EQUAL ACCESS AND SHARED BENEFITS

Opening up access and benefits will require alternative operating and financial arrangements for new and existing facilities. Diversified partnerships for ownership and operations of new facilities and retroactive phase-in for existing facilities could include financial investment, equity stake, and/or rights to output and product. Rules and guidance should be nondiscriminatory and equitably applied with respect to access and benefit. This could be particularly important for creating a framework for regional storage or disposal sites.

With respect to reprocessing, the significant costs of building facilities provide an incentive to states to seek outside investment. While there are probably some legal hurdles to widening international investment in what have traditionally been government-owned operations, there are precedents (e.g., Eurochemic and United Reprocessors Group) for alternatives and there are economic and political incentives on both supplier and recipient sides for widening ownerships.

SHARED COSTS AND BURDENS

Current technology holders should accept the same restrictions as new entrants so that the changes to the system would not be disproportionately felt by nontechnology holders. Applying IAEA safeguards to all existing enrichment and reprocessing plants and the conclusion of Additional Protocols cannot be achieved overnight, but may be possible within ten years, as noted in Figure 5. In addition, all facilities utilizing HEU or Pu (such as MOX fuel fabrication plants) will need to be safeguarded.

In addition to safeguards, all operating enrichment and reprocessing facilities should implement internationally accepted standards and good practices for security and safety. This includes implementation of IAEA INFCIRC-225 and other relevant guidance. Operators can also participate in and implement the recommendations of international industry groups like World Association of Nuclear Operators (WANO) and World Institute for Nuclear Security (WINS). In all cases, the establishment of an independent regulator is an essential element of implementation of required security and safety systems.

In some cases, the additional financial burden of activities required under this best practices approach would be impossible to implement without providing more money. A sustainability "tax"—perhaps scaled to the necessary effort—should be required for all proposed and operating facilities.

EARLY DETECTION

As practiced now, access to sensitive technology is controlled ostensibly for proliferation reasons, but guidelines are set commercially for proprietary reasons. Moreover, there is no uniformity across centrifuge producers. Experts need to improve consistency on how sensitive technology and processes, including manufacture, installation, and maintenance, are controlled through limiting the ability to copy and build components, limiting the ability to modify and operate facilities, and limiting advance planning time. These measures will help ensure that detection is early, credible, and unambiguous, facilitating intervention in case of breakout.

MINIMIZING MATERIAL

To meet the objective outlined above, states must make political commitments and regulatory assurances that there will be no HEU production, no new facilities licensed for HEU production or use, and no civil HEU-fueled facilities. These HEU commitments must be matched with political commitments and regulatory assurances of no plutonium separation unless there is a concurrently available path for consumption. Technical and commercial partnerships to explore ways in which fast reactor research can be oriented toward minimizing HEU use and Pu production should be expanded and politically supported. Countries also need to explore whether reactors, reprocessing, and fuel fabrication activities should be collocated in nuclear "islands," and whether these islands should be under multinational control.

MARKET-BASED EXPANSION

There is no current commercial need for additional uranium enrichment capacity to fuel the existing and near-term operational fleet of nuclear power reactors globally. As a result, market-based expansion is already the foundation of commercial activities for the front end of the fuel cycle. In addition, the international uranium fuel market has functioned effectively for decades. However, uneconomic national enrichment facilities are being built and considered in a number of countries. Any new capacity should meet a market test and only when there is a commercial driver should capacity be expanded.

Regarding spent fuel management and disposal, there is an unmet market need for national, regional, or international joint approaches. Through cooperation, this market demand can be met but requires high-level political drivers based on a national or international security assessment that drives decisionmaking for back-end solutions.

INCENTIVES AND CONSEQUENCES

To establish the norms of behavior that support nuclear security, nonproliferation, and commercial objectives, there must be incentives for those acting within the bounds of acceptability and consequences for those who do not. There are three paths for this: commercial (such as contractual obligations and codes of conduct), national (such as nuclear cooperation agreements), and international (such as treaties or other agreements). These approaches are not mutually exclusive. For example, if a new facility is built or new activities are pursued that are consistent with the best practices, then a country or company can participate freely and fully in the fuel market and gain access to international or regional spent fuel repositories, fuel leasing/take-back arrangements, or other back-end cooperative arrangements. These facilities or activities could benefit from suppliers' provision of advance consent rather than requirements for prior consent. Eventually, in a fully realized system, consent rights could become unnecessary and obsolete.

If a new facility or activity is not judged to be consistent with best practices, there must be consequences. These could include targeted limits on:

- Uranium supply to the country, including other facility inputs such as spent nuclear fuel (SNF)
- Conversion services
- Fuel fabrication services
- Ability to purchase or use the product of fuel cycle facilities
- Ability to send high-level waste or spent fuel to a repository
- Commercial interaction by third parties with any nuclear entity in the country

Taken together, these limits constitute a general ban on nuclear-related international commerce for the country in question.

The judgment of whether activities meet new best practices could be drawn by individual states or commercial entities. Such a responsibility could also be delegated to the IAEA for expression through a statement such as the "broader conclusion" that the IAEA draws regarding states' compliance with safeguards obligations. This evaluation could begin immediately for new facilities and be tied to a defined transition plan for existing facilities (based on national and corporate policies).

Figure 4. Action Plan for Implementation

BEST PRACTICE	STEPS	WHO	HOW	TIME FRAME
EQUAL ACCESS AND SHARED BENEFITS	Diversified partnership for enrichment, reprocessing facilities	Companies (private industry, state-owned, and hybrid)	Examine existing models, define appropriate diversification models	10 years for existing, from beginning for new
	Defined waste path	States, suppliers, and operators	Continue international cooperation, include in supply contracts	10 years for interim approaches, 30 years for long-term disposal
SHARED COSTS AND BURDENS	Safeguards: Additional Protocol and international safeguards on <u>all</u> civilian enrichment and reprocessing (E/R) plants	States, IAEA	Conclude agreements, implement safeguards, amend voluntary offer safeguards agreements for states without comprehensive safeguards	5 years for conclusion of APs, 10 years for implementation of safeguards at E/R plants
	Implementation of INFCIRC-225 (most recent revision)	States	Secure public commitment, embed in legal, regulatory systems	3 years
	Independent regulator	States	Ensure through domestic legal and institutional authorities	5 years
	Sustainability tax	States (via regulators) and IAEA	Define who will need resources for obligations (IAEA, etc.), negotiate arrangements	10 years, and paired to above-stated commitments so that there is no gap
EARLY DETECTION	Black-box and operational separation	New enrichers, technology producers, and operators	Engage in expert-level dialogue, determine applicability to known technology, define appropriate separation of roles	From beginning for new facilities

Figure 4. (cont.)

MINIMIZING WEAPONS-USABLE MATERIAL	No production or use of HEU	States and regulators	Require policy commitments, voluntary reporting, technical R&D for conversion, international cooperative programs for removals	Immediately following policy commitments, phased in as technical approaches are validated
	Drawdown of HEU and Pu stockpiles	Operators, states, and regulators	Secure policy commitments, licensing for utilization of Pu, downblending of HEU	Immediately following policy commitments and technical assessment of facilities; 20–30 years for drawdown of excess inventory
MARKET-DRIVEN EXPANSION	None-already driving commercial activity for enrichment	Companies	Continue limiting new capacity until there is a market need	Current and ongoing
	Development of regional and international options to meet market demand for the back end	States, regulators, and companies	Create consortia, explore technical and siting options, define collective needs	Beginning immediately for option development; 10–30 years for implementation
INCENTIVES AND CONSE-QUENCES	Requiring prior consent for processing of supplier's material and/or equipment	States and suppliers	Include in nuclear cooperation agreements and standard supply contracts; provide opportunities for material uses when consistent with the best practices	Beginning immediately for new agreements and phased in over 15 years for existing as they are renewed

Such an approach would counter the current de facto monopoly of the fuel cycle owner-states and increase access to the peaceful benefits of nuclear energy through a layered system that includes the active participation of states in facilities and fuel assurances. Current technology holders would have to accept new partners and regulatory structures, so the changes to the system would not be disproportionately felt by nontechnology holders. In order to make the decision to support a change easier for a state-owned entity or private company, new approaches to the operation of fuel cycle facilities should also address concerns about energy security and nondiscrimination.

Many of these measures are already implemented by many countries. Encouraging some key states to take action will result in improvements even if the complete vision is not fully realized. Figure 5 indicates where a few sample countries stand with respect to these measures.

Building Support for a New Approach

Virtually all proposals to reform the fuel cycle have floundered in the face of political, economic, commercial, and legal concerns. Mostly, they failed to create the necessary buy-in because they lacked the ability to:

1. Preserve the robustness of the nonproliferation regime and enhance the international nonproliferation consensus.
2. Avoid disruption of the fuel market.
3. Cultivate a shared understanding among industry actors and governments of the deficits in the existing system and the need for change.
4. Apply to existing good-faith actors as well as preempt and expose potentially destabilizing and dangerous actions.
5. Provide multiple options for secure access to peaceful uses of nuclear technology and create nondiscriminatory access to fuel cycle products and services.
6. Link the front and back ends of the fuel cycle to the needs and interests of users.
7. Provide a path for transitioning from the current system to a new one, with practical steps for operators, international organizations (including the IAEA), governments, and regulators.
8. Compare favorably to the status quo, not the ideal, and address both existing facilities and stocks of materials and future facilities and material production.

The approach outlined in this report has attempted to meet those guidelines and thus improve the potential for political and commercial drivers to spur change rather than impede it.

Proposals are more likely to lead to real change if they are consistent with an effectively functioning market and create incentives to participate. Not every state is starting in the

Figure 5. State of Play on the Action Plan for Sample Countries

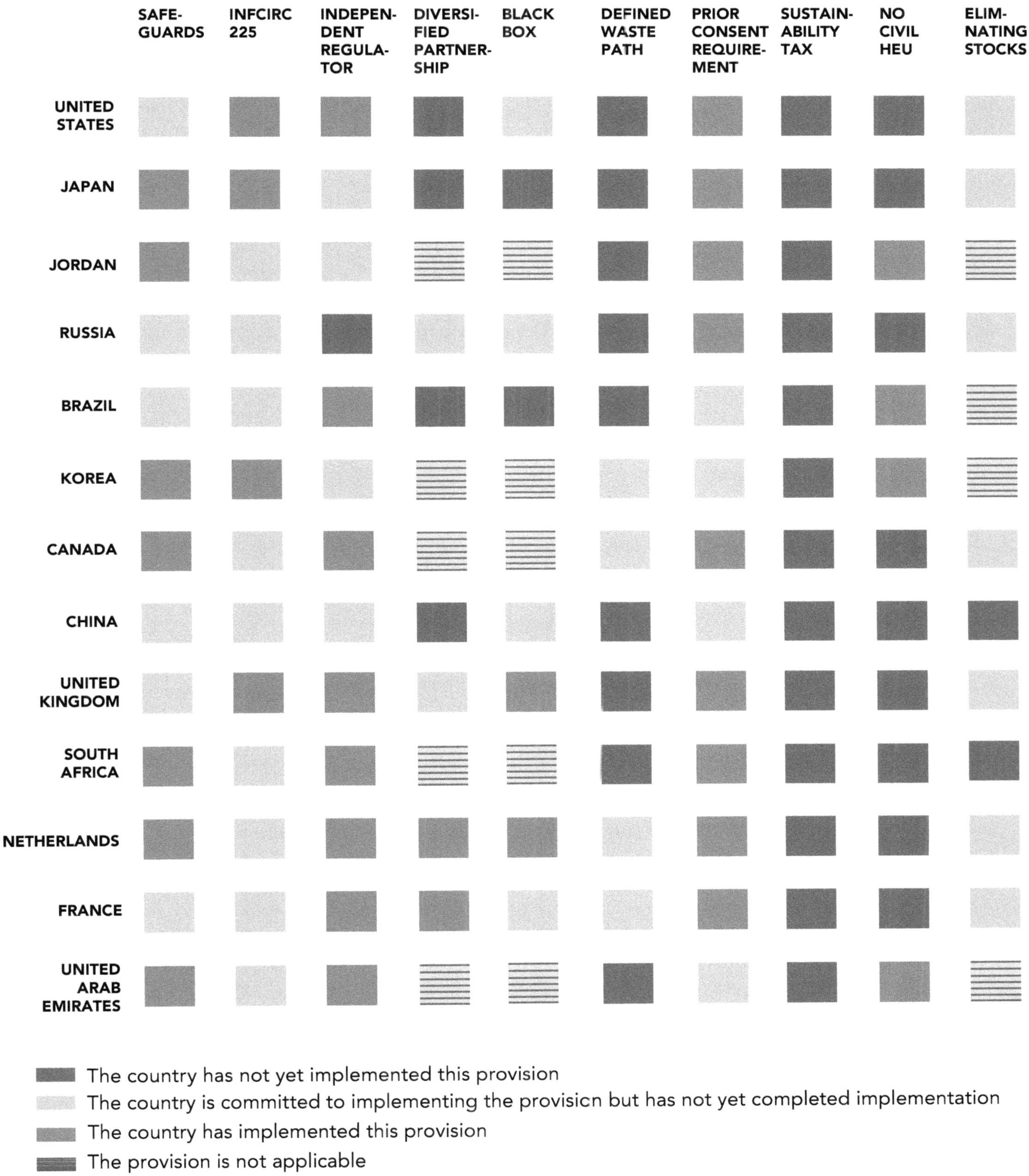

same place, even with regard to the measures that could be implemented without fundamental changes to the existing system.

Ultimately, an integrated fuel cycle approach should avoid interfering with what already exists and works, leverage existing trends and positive dynamics, and add options for addressing remaining problems. It needs to reduce risks, possibly by constraining actions, and provide the information and access necessary for quicker responses against a proliferator or otherwise noncompliant state.

Building support for a best practices approach to the nuclear fuel cycle could start with a Track 1.5 dialogue focusing on "Sustainability and Security for the Nuclear Fuel Cycle" that adds to the work of several organizations in the nongovernmental organization community by engaging more specifically on key gaps in concepts and approaches. A targeted discussion among government, industry, and stakeholders (including international organizations and the policy community) could focus on the following:

- Black-box approaches to sensitive fuel cycle facilities, including work to rationalize approaches designed to protect industry secrets and those designed to prevent proliferation of sensitive technologies
- Fuel cycle decisionmaking in countries pursuing advanced nuclear fuel cycles (in contrast to technical discussions), particularly in Russia, China, and India
- Security and safeguards requirements for older spent nuclear fuel, particularly low burn-up fuel
- The technical and political desirability of collocation of reactors, reprocessing, and fuel fabrication in nuclear islands, potentially under multinational control
- The applicability of lessons of the European spent fuel management experience in other regions, particularly Asia and the Middle East
- Perceptions by governments and industry on the security of supply and on methodologies for determining market need for fuel cycle services.
- Developing consequences for destabilizing behavior

Existing fora, such as the International Forum for Nuclear Energy Cooperation, the Nuclear Suppliers Group, the Asia-Pacific Safeguards Network, or the International Atomic Energy Agency, could address these topics but it will be important to engage industry fully in this process. To this end, a standing joint industry-policy community forum for discussion of security and proliferation concerns related to the nuclear fuel cycle should be initiated. Such a forum could help draft, for example, fuel cycle industry principles of action that would support a new best practices approach.

APPENDIX: NUCLEAR FUEL CYCLE PRIMER

It is likely that a "friends of sustainable nuclear energy" group that is widely representative of the governments and industries of supplier and recipient countries will be necessary to help cement a more sustainable approach to the nuclear fuel cycle in the future.

THE NUCLEAR FUEL CYCLE (SHOWN IN FIGURE A-1) IS COMMONLY DIVIDED INTO TWO STAGES: THE FRONT END, IN WHICH URANIUM IS PROCESSED THROUGH CONVERSION AND ENRICHMENT PLANTS AND THEN FABRICATED INTO FUEL FOR REACTORS; AND THE BACK END, IN WHICH FUEL IS HANDLED AFTER IRRADIATION IN A REACTOR. Reactors are generally considered the operations process between the front and back ends.

FRONT END

- *Uranium Exploration, Mining, and Milling:* Uranium ore is mined and processed to produce uranium ore concentrate, commonly called U_3O_8 or yellowcake.
- *Conversion:* The concentrate is converted to uranium hexafluoride gas (UF_6).
- *Enrichment:* The UF_6 is enriched to contain up to 5 percent of the isotope U-235 in the enriched uranium product.
- *Fuel Fabrication:* The enriched uranium product is processed into uranium dioxide (UO_2) powder and pressed into pellets. The pellets are then encased in metal tubes to form fuel rods, which are arranged into a fuel assembly ready for introduction into a reactor.

BACK END

- *Spent Fuel Interim Storage:* Spent fuel removed from the reactor is temporarily stored in spent fuel pools, which helps to dissipate heat and radioactivity. It is then sometimes removed to massive air-cooled dry casks for further storage.

Figure A-1. A Nuclear Fuel Cycle Process

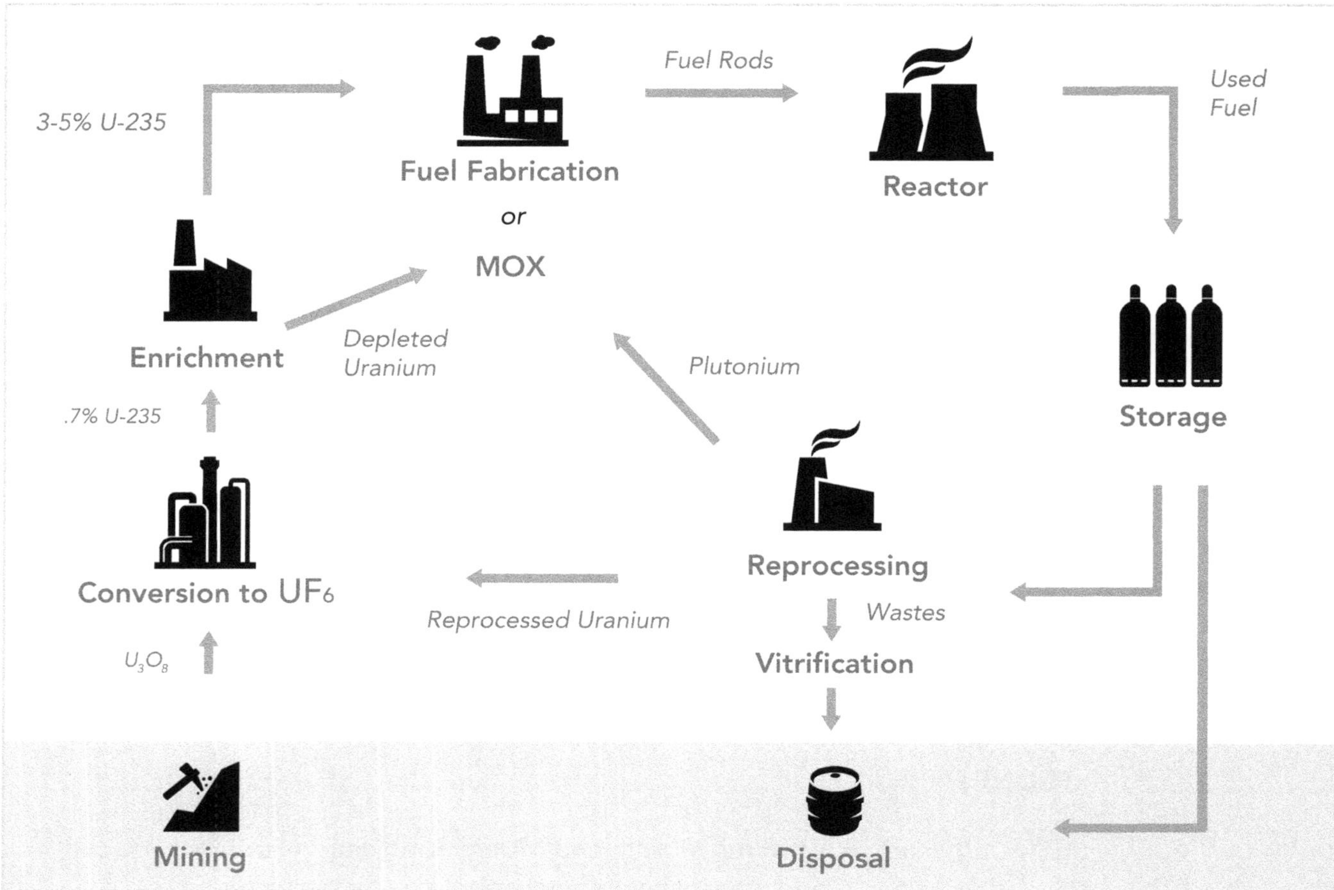

- *Reprocessing:* Fuel can eventually be taken to a reprocessing facility, where it is chemically separated into its three components: uranium, plutonium, and waste that contains fission products.

  Reprocessing is sometimes referred to as recycling, but this report uses the term "recycling" to encompass the reuse of this material in new fuel. It is possible to reprocess without recycling the material back into reactor fuel, such as "conditioning" spent fuel for long-term storage by partitioning especially long-lived fission products like cesium-137 and strontium-90.

- *Recycling:* The uranium from reprocessing, which typically contains a slightly higher concentration of U-235 than occurs in nature, can be reused (recycled) as fuel after conversion and enrichment. The plutonium can be directly made into mixed oxide (MOX) fuel, in which uranium and plutonium oxides are combined. In fast neutron spectrum reactors, both plutonium and fission products can be consumed in the reactor.

Uranium mining and milling is the first stage in the fuel cycle and although subject to some regulations, is generally outside the IAEA safeguards process. The traditional starting point of safeguards is after conversion of uranium oxide into uranium hexafluoride—the point at which the uranium product is ready to be enriched.

- *High-level Waste Disposal:* Spent fuel and other high-level wastes are expected to be placed in a permanent geological repository (none are yet built).

The decision to reprocess and/or recycle spent fuel determines whether a nuclear program operates an "open" (once-through) or "closed" fuel cycle. In the open cycle, nuclear fuel is used once, and the spent fuel is disposed of in either an interim storage facility or (once one is built) a permanent geological repository. The majority of countries have opted for this approach, although none has yet opened a permanent repository. A few countries have chosen a closed fuel cycle, reprocessing the spent fuel and recycling it as reactor fuel. Until now, they have only been able to reuse a relatively small amount of fuel in conventional or advanced reactors; research and development of multiple recycling of fuel for fast reactors is under way.

Uranium Production

Uranium mining and milling are concentrated in the countries with the largest resources, many of which do not operate nuclear power plants. Three countries accounted for just under 64 percent of global uranium production in 2012. Kazakhstan was the world's top producer with 36.5 percent of primary production. Canada was second with 15 percent, followed by Australia with 12 percent, Namibia and Niger with 8 percent each, and Russia with 5 percent. According to the World Nuclear Association, global production increased by roughly 5,000 metric tons in 2012, due largely to increased output in Kazakhstan, Namibia, and Australia. The 10 top-producing countries were responsible for virtually all of this output; eight companies were responsible for 88 percent of global mine production; and 64 percent of primary production came from the 15 most productive sites.[1]

Major uranium producers include private and government-owned entities, and there are many partnerships, joint ventures, and other investment or ownership arrangements.[2] Some of the larger uranium producers such as Cameco, Areva, and Rosatom are also active in other parts of the fuel cycle.[3]

Uranium is not a publicly traded good, unlike other commodities and precious metals. While 90 percent or more of the trade in uranium has historically been through long-term contracts, these contracts often reference a spot price, which is determined by one or more prices published by specialist broker/traders who closely monitor bids and offers. Long-term contracts are of varying duration, from five to 10 or even 15 years; they require the mining company to meet production specifications (such as purity and isotopic concentration) and transport milled uranium to a conversion facility by a pre-agreed date. Utilities often enter into several of these contracts with various suppliers to ensure security of supply.[4] Utilities also purchase uranium on the spot market to take advantage of a drop in price.

Conversion

Conversion of milled uranium to uranium hexafluoride, a precursor to enrichment, has been hampered by low to nonexistent profit margins, resulting in outdated facilities with safety concerns. The five principal conversion companies—Areva, Rosatom, Cameco, Westinghouse, and ConverDyn—are global service providers, meaning they provide conversion services for customers in a variety of markets. Most conversion contracts consist of

---

1. World Nuclear Association, "World Uranium Mining Production," October 2014, www.world-nuclear.org/info/Nuclear-Fuel-Cycle/Mining-of-Uranium/World-Uranium-Mining-Production/#.Ue_hz9JgS8A.

2. For example, Uranium One is 51 percent owned by Atomredmetzoloto, and all of Atomredmetzoloto's nominal equity stakes in Kazakh production have been transferred to Uranium One, a Canadian-based, publicly traded company that also has uranium assets in the United States, Australia, and Tanzania. The Kazakh operation Tortkuduk, now the largest in situ recovery uranium project in the world, is being exploited by an Areva-Kazatomprom joint venture called Katco. British-based Rio Tinto, the world's fourth largest uranium producer by company (after Cameco, Kazatomprom, and Areva), is majority owner of Energy Resources of Australia, operator of the Ranger open-pit mine, which accounted for 6 percent of global output in 2010; Rio's own Rossing open-pit mine in Namibia also accounted for 6 percent. China, Russia, and South Korea have been actively pursuing equity stakes in uranium projects, particularly in Canada and southern Africa. Mongolia and Tanzania are also attracting interest.

3. For example, Cameco owns and operates a uranium refinery, a uranium conversion plant, and fuel manufacturing operations. Both Areva and Rosatom are involved in virtually every stage of the fuel cycle through joint ventures in other countries and wholly owned operations in multiple locations. Rosatom's involvement in uranium production is through a wholly owned subsidiary of Atomredmetzoloto.

4. Technical and political bottlenecks to the trade in uranium, particularly to China and India, are evident, and suppliers and recipients are searching for ways to circumvent them. The sheer quantity of material heading east by rail at that border—together with the differently gauged railways that make for a logistical nightmare—means that there are considerable backups at the Chinese border town of Alashankou in the Xinjiang Autonomous Regions. Until recently Cameco engineered complex "flag swaps" at its Canadian conversion facilities, where it switched Canadian-origin uranium with concentrates from other countries to China, because it was not allowed to ship direct to China. Moreover, the Kazakh government controls access rights and is not quick to grant the privilege.

long-term agreements between nuclear power plant operators and these major firms. Areva and Rosatom, however, are vertically integrated companies and encourage their customers to buy a range of services, including uranium, conversion, enrichment, and fabrication.

Demand for conversion services has been hurt by the availability of secondary uranium sources, including not only spot market material, but Russian highly enriched uranium down-blended under the U.S.-Russian Megatons to Megawatts program. Currently, conversion capacity exceeds demand. For companies like Areva, a capability in conversion may simply be the necessary cost of providing full cycle fuel services and supporting more profitable areas of the cycle.

Enrichment

Six companies commercially enrich uranium, though the market is highly concentrated among five providers—Urenco, Areva, Eurodif (59 percent owned by Areva), the United States Enrichment Corporation (USEC), Russia's Tenex, and the China National Nuclear Corporation.[5]

Most uranium is enriched under long-term contracts (five years or more), with some enrichment traded in a spot market. The movement toward centrifuge technology is likely to reinforce this trend, as centrifuge capacity is generally added only upon securing long-term enrichment contracts. Suppliers therefore have as much interest in long-term contracts as recipients.

Trade policy also plays a major role in enrichment markets. In the European Union, an informal quota system imposed by Euratom limits the supply of Russian enriched uranium product to approximately 20 percent of all enriched uranium product used in order to protect European producers (Urenco and Eurodif). In 2008, the United States and Russian signed an agreement allowing U.S. utilities to import enriched uranium directly from Russia. The compromise allowed for small amounts of imports until 2013, after which an import quota mechanism is in effect up to 2020.

Centrifuge technology has transformed the enrichment landscape over the last few decades. The United States, once the dominant provider, is in danger of disappearing from this market altogether. Because of the high entry costs and sensitive nature of the technology, private-sector investment in enrichment has been limited. At present, only General Electric is looking at entering the enrichment business; it has partnered with Hitachi of

---

5. Japan Nuclear Fuel Ltd., as well as India, Pakistan, Brazil, Argentina, Iran, and North Korea, all have enrichment capacities but minimal impact on commercial markets. The China National Nuclear Corporation's June 2013 announcement that it had mastered enrichment using domestic gas centrifuge technology and intended to meet Chinese demand will affect the market in the medium term should the China National Nuclear Corporation prove able to deliver. Zhang Xiaobo, "China Develops Own Tech to Enrich Uranium," *Global Times*, June 25, 2013, www.globaltimes.cn/content/791301.shtml#.Ue_q7NJgS8C.

Japan and Cameco of Canada to form Global Laser Enrichment, which is attempting to develop a laser enrichment technology called Silex. If Global Laser Enrichment succeeds in commercially developing the technology, which might be possible as early as 2015, it could provide further efficiencies and cost reductions. However, laser enrichment facilities can have a small physical footprint, which could make them even more difficult to detect than centrifuge facilities and pose a greater proliferation risk as a result.

Fuel Fabrication

Fuel fabrication is a more specialized engineering process than mining, conversion, and enrichment because fuel rods must be tailored to the specific needs of each reactor. The dimensions of the fuel pellets and other components of the fuel assembly are precisely controlled to ensure consistency in the fuel characteristics.

Reactor vendors or their affiliates are the major fuel fabricators. Most fuel contracts require the fuel fabricator to provide sufficient information to develop a detailed fuel design for the reactor, a process that takes about five years and involves a conceptual design followed by years of testing and then at least a year's effort to acquire a license. As the industry has consolidated over the past three decades, so too have fuel fabricators. There are now roughly 11 in the world, including Areva in France and the United States, the China National Nuclear Corporation in China, Korea Nuclear Fuel, Rosatom's TVEL in Russia, and General Electric and Westinghouse in the United States. Consolidation has also led to vendors branching out from their traditional markets. Utilities tend to favor contracts with shorter terms than those for enrichment services in order to accommodate small but frequent fuel design upgrades that increase reliability, although they maintain long-term relationships with suppliers.

Reprocessing

The only commercial method for reprocessing used at the moment is the multistage plutonium–uranium extraction process, which uses nitric acid to dissolve the fuel elements. So far, almost 90,000 metric tons out of 290,000 metric tons of spent fuel discharged from commercial power reactors has been reprocessed. Reprocessing capacity is now some 4,000 metric tons per year for normal oxide fuels, but not all of it is operational. Between now and 2030 some 400,000 metric tons of used fuel is expected to be generated worldwide, including 60,000 metric tons in North America and 69,000 metric tons in Europe.[6] Uranium commands a much larger share of the reprocessing market than MOX fuel, but a relatively small share of the uranium market overall. The World Nuclear Association estimates that MOX fuel accounts for only 2 percent of new nuclear fuel used globally. Unlike the plutonium trade, the reprocessed uranium market is not governed by stringent legal restrictions and is thus

6. World Nuclear Association, "Processing of Used Nuclear Fuel," September 2014, www.world-nuclear.org/info/inf69.html.

open to greater commercial involvement from both a supply and demand standpoint. Commercial nuclear utilities retain ownership of all of the materials produced during recycling and reprocessing, including wastes. While they can decide whether to opt for reprocessing, they can't as a rule sell their fuel to another organization for reprocessing.

Reprocessing contracts are long term in part because of the length of time between the discharge of spent fuel assemblies and fabrication of new fuel, but also because of the enormous costs and planning associated with reprocessing "campaigns" at the chemical separation facilities. The utilities also must agree to take back the high-level wastes arising from reprocessing activities, although these often are stored at reprocessing facilities for years before they are returned. Reprocessors have to pay the costs of disposal for irradiated equipment and other items associated with the operations themselves, all of which have limited shelf lives; it is not practical to pass these costs on to customers.

The United Kingdom, France, and Russia currently separate the largest amounts of plutonium, and both the United Kingdom and France temporarily store plutonium for other countries. As of January 2012, a total of 260 metric tons of plutonium had been declared by the United Kingdom, France, Russia, and Japan through their reports under IAEA INFCIRC-549. An estimated two metric tons of additional material is stored in Germany as MOX fuel. The United Kingdom accumulates the largest amounts of civilian plutonium each year, although that rate is decreasing because of problems at its Thermal Oxide Reprocessing Plant at Sellafield in Cumbria, which is scheduled to close in 2018 when all existing reprocessing contracts are completed.[7] France's stockpile has decreased slightly since 2012 primarily because Areva has been shipping it back to foreign customers that have terminated their reprocessing contracts. Russia continues to separate plutonium at about one to two metric tons per year and plans to use its excess plutonium for a commercial breeder program starting in 2020.[8] Japan is mulling its options for the use of MOX fuel in its light water reactors, given the present difficulty it finds itself in with no operating reactors.

Areva leads the international market for reprocessing, but has lost most of its foreign customers, with the result that it is operating at roughly half capacity. Areva has only one ostensible competitor, the United Kingdom, and its reprocessing capacity and plans for managing its plutonium stockpiles remains unsettled. In 2011 the UK Nuclear Decommissioning Authority (NDA) closed the Sellafield MOX plant, Britain's only commercial MOX fuel fabrication facility. However, in December 2011, the UK Department of Energy and Climate Change announced "a preliminary policy view to pursue reuse of plutonium as mixed oxide fuel" in order to convert Britain's large stockpile of separated plutonium into fuel.[9] As of

7. "Sellafield Thorp Site to Close in 2018," BBC News, June 7, 2012, http://www.bbc.com/news/uk-england-cumbria-18353122.

8. International Panel on Fissile Materials (IPFM), *Global Fissile Material Report 2011: Nuclear Weapon and Fissile Material Stockpiles and Production* (Princeton, NJ: IPFM, 2011), http://fissilematerials.org/library/gfmr11.pdf; Gary Peach, "Russia: Rosatom Sticks to 2020 Deadline for Breeder Launch," *Nuclear Intelligence Weekly*, March 9, 2012.

9. UK Department of Energy and Climate Change, "Charles Hendry Written Ministerial Statement on Nuclear Energy Matters," December 1, 2011, https://www.gov.uk/government/news/charles-hendry-written-ministerial-statement-on-nuclear

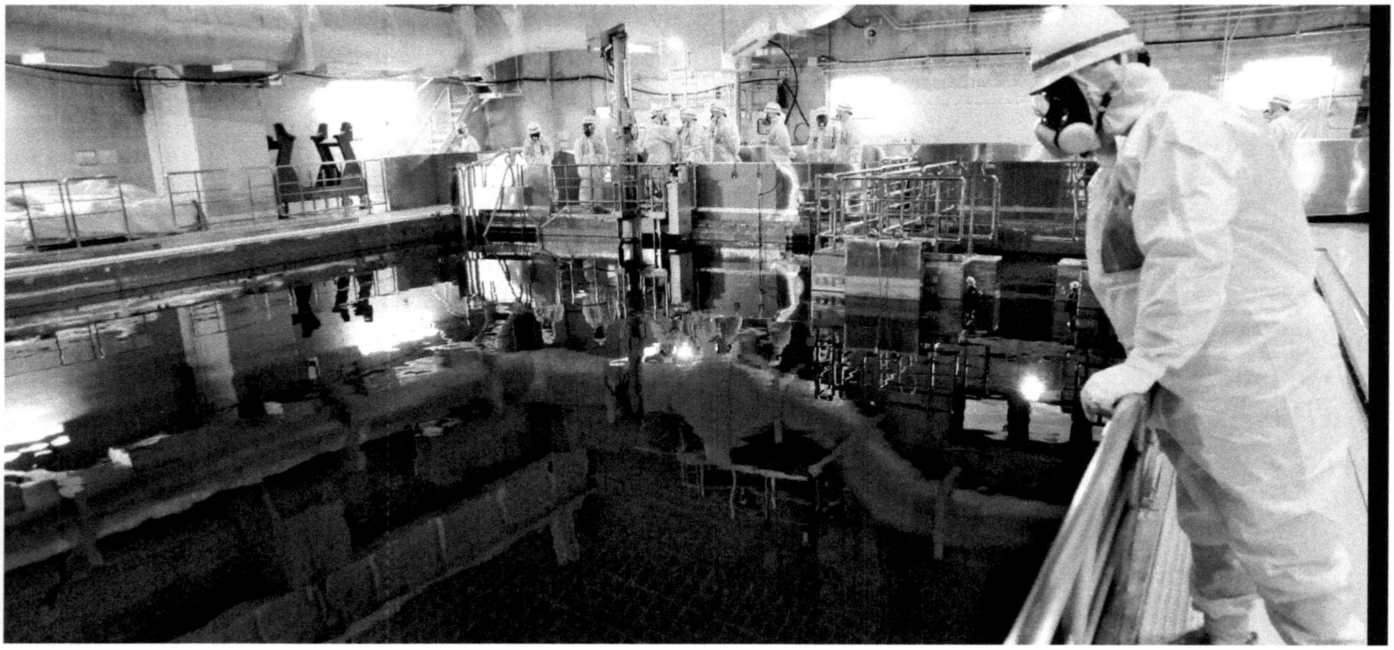

At most reactors, spent nuclear fuel is cooled for a period of time (often at least five years) before it is moved either to a central spent fuel pool or dry cask storage. The risks of pools at reactors was highlighted in the Fukushima Daiichi nuclear power plant accident in March 2011. Some kinds of configurations for these pools can pose risks that terrorists can exploit.

February 2012, the NDA sought proposals for "potential alternative approaches for managing the UK's plutonium stocks" while continuing to support the government as "it progresses its preferred policy of converting the material into mixed oxide fuel (MOX) for reactors."[10]

Storage of Spent Nuclear Fuel

All nuclear power programs require a disposal pathway for the resulting spent fuel and/or high-level radioactive waste. Geological disposal is the internationally accepted strategy for permanently isolating this waste from the accessible environment, but no repositories for commercial waste are currently operating.

As of the end of 2012, about 270,000 metric tons of spent fuel (in the form of heavy metal) was in storage worldwide, most of it at reactor sites.[11] About 90 percent was in

-energy-matters.

10. UK Nuclear Decommissioning Authority, "Proposals Sought for Alternatives to Re-Use of Plutonium as MOX Fuel," February 23, 2012, http://www.nda.gov.uk/2012/02/proposals-sought-for-alternatives-to-re-use-of-plutonium-as-mox-fuel/.

11. World Nuclear Association, "Radioactive Waste Management," September 2014, www.world-nuclear.org/info/Nuclear-Fuel-Cycle/Nuclear-Wastes/Radioactive-Waste-Management/#.UfAOmtJgS8A.

Retrievability in Spent Fuel Repositories

> "Retrieval is always possible in principle. Engineering methods to allow retrievability are available, even though they become more complex and expensive as the step-wise closure of the repository progresses and with increasing time after closure of the repository."[a]

According to the Blue Ribbon Commission on America's Nuclear Future, retrievability could be considered a desirable or necessary feature of facility design for two main reasons: monitoring the nuclear waste, to help confirm the behavior of the repository and, if necessary, allow for removal of the waste; and preserving the option of retrieving spent fuel for future reprocessing and recycling. The Commission supported U.S. retrievability requirements, noting that they are intended to make it possible to remove waste in case of problems, not to enable recovery and reuse of the material.

Retrievability should be seen in the context of all other considerations, including what the volunteer community wants as well as engineering considerations. If retrievability is imposed without regard to other factors, the system design could become quite complex. Where retrievability may not be desirable or technically feasible (for example, in the case of borehole disposal), other forms of assurance, such as a pilot facility or processing and waste tailoring options that would minimize the need for retrieval, may be appropriate.[b] However, it will be decades before geological repositories begin accepting waste, and they will operate for decades and might be sealed only after protracted monitoring phases. In any event,

> Whatever we decide now, there is no compulsion whatsoever for eventual operators and regulators of a repository to adopt our philosophy or respond as we do to present-day drivers. Thus there will be considerable opportunity for changes in approach to decision-making before a repository has reached the end of its operational life. What does remain our responsibility is to ensure that future operators can complete the task safely, perhaps with their own changes, and certainly in their own time, rather than leaving them with an incompletely designed facility that is not intrinsically safe at all times, both operational and post-closure.[c]

a. Neil Chapman and Charles McCombie, *Principles and Standards for the Disposal of Long-Lived Radioactive Wastes* (Oxford: Elsevier, 2003), 66.

b. Blue Ribbon Commission on America's Nuclear Future, *Report to the Secretary of Energy* (Washington, DC: Blue Ribbon Commission on America's Nuclear Future, January 2012).

c. Chapman and McCombie, *Principles and Standards for the Disposal of Long-Lived Radioactive Wastes*, 67.

storage ponds and the balance in dry casks. Every year, nuclear power reactors across the world generate about 10,500 metric tons of heavy metal. Of that, roughly 8,500 metric tons goes into long-term storage and about 2,000 metric tons is allocated for reprocessing, but much of that remains in interim storage.[12] The 10 largest nuclear power generators produce about 87 percent of these totals. None of these has yet opened a commercial nuclear waste repository.

Spent fuel pools were never designed to be a long-term storage solution. Many sites have had to install high-density racks and/or add dry storage casks on site. Storage is limited by the size of the pool and criticality-control geometry, and acquisition of dry casks telegraphs a rather long-term commitment to on-site storage, often raising fairness and political issues.[13] Taiwan faces an acute space shortage in the spent fuel pool at its Chinshan plant, South Korea will face the same problem at its Kori plant in several years, and Japan will also confront this problem at many plants in the coming years.

Most countries with nuclear programs have opted for direct disposal in a permanent repository, but political and technical difficulties have prevented them from actually building one. There is also some debate concerning the merits of retrievable versus nonretrievable repository designs, based in part on perceptions of whether spent fuel is a resource or a waste. Finland and Sweden have made more progress than any other country; they have selected repository sites and begun licensing and other activities.

Siting and building a repository is expensive: According to a 2008 U.S. Department of Energy life cycle cost estimate, it would cost US$96.2 billion (in 2007 values) to license, construct, operate, and close a repository at Yucca Mountain of sufficient size to dispose of 122,000 metric tons of commercial and defense spent nuclear fuel and high-level waste. The cost share assigned to 109,000 metric tons of commercial waste (80 percent of the total) was US$77 billion. In 2010, the Swedish Nuclear Fuel and Waste Management Company estimated that the total cost for the Swedish repository would be 123 billion Swedish kronor (approximately US$22.6 billion), although this figure was revised upward in 2011. The total cost of the Finnish repository is approximately €3 billion—roughly €650 million in investment costs, €2.1 billion in operating costs through 2118, and €250 million in decommissioning and closure costs.[14]

---

12. Harold Feiveson et al., eds., *Spent Fuel from Nuclear Power Reactors: An Overview of a New Study by the International Panel on Fissile Materials* (Princeton, NJ: IPFM, June 2011), www.fissilematerials.org/ipfm/site_down/ipfm-spent-fuel-overview-june-2011.pdf.

13. An expedient interim solution, particularly in the United States, is to create more on-site storage space. However, this is not always easy. For example, in South Korea, where nuclear plants are close to the surrounding communities, the local community must be consulted before any on-site storage expansion can take place.

14. Blue Ribbon Commission on America's Nuclear Future, *Disposal Subcommittee Report to the Full Commission— Updated Report* (Washington, DC: Blue Ribbon Commission on America's Nuclear Future, January 2012), http://cybercemetery.unt.edu/archive/brc/20120620220845/http://brc.gov/sites/default/files/documents/disposal_report_updated_final.pdf; World Nuclear Association, "Nuclear Power in Sweden," October 2014, www.world-nuclear.org/info/inf42.html; Posiva Oy, "Total Costs and Funding for Final Disposal," n.d., http://posiva.fi/en/final_disposal#.VEalLvldVyx.

ABOUT CSIS

For over 50 years, the Center for Strategic and International Studies (CSIS) has worked to develop solutions to the world's greatest policy challenges. Today, CSIS scholars are providing strategic insights and bipartisan policy solutions to help decisionmakers chart a course toward a better world. CSIS is a nonprofit organization headquartered in Washington, DC. The Center's 220 full-time staff and large network of affiliated scholars conduct research and analysis and develop policy initiatives that look into the future and anticipate change.

Founded at the height of the Cold War by David M. Abshire and Admiral Arleigh Burke, CSIS was dedicated to finding ways to sustain American prominence and prosperity as a force for good in the world. Since 1962, CSIS has become one of the world's preeminent international institutions focused on defense and security; regional stability; and transnational challenges ranging from energy and climate to global health and economic integration.

Former U.S. senator Sam Nunn has chaired the CSIS Board of Trustees since 1999. Former deputy secretary of defense John J. Hamre became the Center's president and chief executive officer in 2000.

ABOUT NTI

The Nuclear Threat Initiative (NTI) is a non-profit, non-partisan organization with a mission to strengthen global security by reducing the risk of use and preventing the spread of nuclear, biological, and chemical weapons and to work to build the trust, transparency, and security that are preconditions to the ultimate fulfillment of the Non-Proliferation Treaty's goals and ambitions.

Founded in 2001 by former U.S. Senator Sam Nunn and CNN founder Ted Turner, NTI is guided by a prestigious, international board of directors. Joan Rohlfing serves as president.

ABOUT THE AUTHORS

KELSEY HARTIGAN is a program officer at the Nuclear Threat Initiative (NTI). Prior to joining NTI, Hartigan worked as a nonproliferation and defense policy analyst for the National Security Network and held positions with the United States Institute of Peace, the Stimson Center and the U.S. Department of State. She is a member of the Institute of Nuclear Materials Management, the International Network of Emerging Nuclear Specialists and Women in International Security. Hartigan is a graduate of Purdue University.

COREY HINDERSTEIN is vice president of international programs at NTI. Prior to joining NTI, Hinderstein was deputy director and senior analyst at the Institute for Science and International Security (ISIS). She edited *Cultivating Confidence: Verification, Monitoring, and Enforcement for a World Free of Nuclear Weapons* (Nuclear Threat Initiative, 2010), which explores the key issues associated with verifying, monitoring, and enforcing the steps needed to move toward a world without nuclear weapons and offers some possible solutions to these challenges. Hinderstein is a member-at-large on the boards of directors of ISIS and WINS, vice president of the international executive committee for the Institute of Nuclear Materials Management (INMM), and member of Women in International Security. Hinderstein holds a bachelor's degree in government with a concentration in international relations from Clark University in Worcester, MA, where she was elected to Phi Beta Kappa.

ANDREW NEWMAN, PH.D., is a senior program officer at NTI focusing on nuclear energy, nuclear waste and nonproliferation. Prior to joining NTI, Newman was a research associate with Harvard University's Project on Managing the Atom. Before that, Newman spent three years with the Nuclear Science and Technology Office at the Australian Embassy in Washington DC and with the Partnership for Global Security in Washington, DC. and in the Office of the Emergency Services Commissioner, Department of Justice, Victoria, Australia. Newman is also an adjunct research associate at Monash University, Victoria Australia where he holds a PhD in political science. Newman co-edited *Japan, Australia and Asia-Pacific Security* (Routledge, 2006).

SHARON SQUASSONI is director and senior fellow of the Proliferation Prevention Program at the Center for Strategic and International Studies (CSIS). She joined CSIS from the Carnegie Endowment for International Peace, where she was a senior associate in the Nuclear Nonproliferation Program. From 2002 to 2007, Squassoni advised Congress as a senior specialist in weapons of mass destruction at the Congressional Research Service (CRS), Library of Congress. From 1992 to 2001, Squassoni held positions in the U.S. Department of State and in the Arms Control and Disarmament Agency. She holds degrees in political science and public management from the State University of New York and the University of Maryland, and a master's in national security strategy from the National War College. She is on the Science and Security Board of the Bulletin of Atomic Scientists and the Board of the Center for Arms Control and Nonproliferation.

www.ingramcontent.com/pod-product-compliance
Ingram Content Group UK Ltd.
Pitfield, Milton Keynes, MK11 3LW, UK
UKHW051418150926
475561UK00005B/239

9 781442 240537